£6-50 ✗

**ELECTRONICS
SERVICING
VOL. 3**

This ................... on or before .................... below.

# ELECTRONICS SERVICING VOL. 3
# Core Studies: Principles and Electronic Circuits

K. J. Bohlman
I.Eng., F.S.E.R.T., A.M.Inst.E.

Dickson Price Publishers Ltd
Hawthorn House
Bowdell Lane
Brookland,
Romney Marsh
Kent TN29 9RW

Dickson Price Publishers Ltd
Hawthorn House
Bowdell Lane
Brookland
Romney Marsh
Kent TN29 9RW


First Published 1982
© K. J. Bohlman 1982
Second Edition 1989

British Library Cataloguing in Publication Data

Bohlman, K. J. (Kenneth John)
  Electronics servicing. – 2nd ed.
  Vol. 3, Core studies: principles and
  electronic circuits.
  1. Electronic equipment
  I. Title    II. Electronics servicing 224 course
  621.381

ISBN 0-85380-196-7

Photoset by
R. H. Services, Welwyn, Hertfordshire
Printed and bound in Great Britain by
Biddles Limited, Guildford and King's Lynn.

# CONTENTS

## Other Books of Interest

**Inspection Copies**

Lecturers wishing to examine any of these books should write to the
publishers requesting an inspection copy.

Complete Catalogue available on request.

# AUTHOR'S NOTE

This book together with Vol. 2 (Principles and Electronic Devices) will fully cover the Core Studies syllabus for Part II of the City and Guilds Electronic Servicing 224 Course.

# CHAPTER ONE

# POWER SUPPLIES

In *Electronic Systems* the purpose of a rectifier was discussed and in *Core Studies* (*Electronic Devices*) the operation and construction of p-n diode rectifiers were explained together with a description of a simple half-wave rectifier. It is now necessary to give further details of the half-wave rectifier circuit and to see how other rectifier circuits can be formed using p-n diode rectifiers so that a.c. can be converted into a suitable d.c. for the operation of electronic equipment.

### Half-wave Rectifier

The basic circuit of a half-wave rectifier is shown in Fig.1.1. It consists of a p-n diode in series with a load (the equipment to be supplied), represented by the resistor R. The a.c. voltage to be rectified is fed between terminals A and B. This input voltage may come directly from the mains supply or from the secondary winding of a mains transformer.

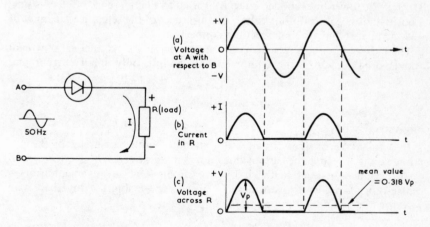

*Fig. 1.1 Half-wave rectifier circuit.*

Considering waveform (a) when terminal A is positive to terminal B the p-n diode will be in a forward biased state and current will flow in the direction shown through the diode and load. During the period when terminal A is negative to terminal B the p-n diode will be in a reverse biased state and no current (or a very small leakage current) will flow in the circuit. As waveform (b) shows, current flows in the load only during the positive half-cycles of the input. If the voltage drop across the rectifier is neglected, the current flowing in the load will be proportional to the input voltage. Thus, with a sinewave voltage input the current in the load will consist of half sinewaves (assuming a perfect rectifier). This current flowing in R will cause half sinewaves of voltage across it; see waveform (c). The voltage across R represents the output voltage of the rectifier and since it is always in one direction it is d.c.

Because there is no output during the negative half-cycles of the input, the mean value of the d.c. output is small. Now, the mean value of a half sine-wave is $0.637 \times$ Peak Value but since the output across R consists of a half sinewave of voltage followed by an equal period of zero voltage the mean value will be:

$$\frac{0.637 \times \text{Peak Value}}{2} = 0.318 \times \text{Peak Value}$$

With a 240V r.m.s. mains input to terminals A and B, the mean d.c. output voltage (neglecting the voltage drop across the rectifier) would be:

$$0.318 \times Vp$$
$$= 0.318 \times 240 \times 1.414V$$
$$\simeq 108V$$

It will be noted that the polarity of the output voltage is such that the upper end of R is positive with respect to the lower end. If, however, the connections to the rectifier are interchanged, the polarity of the voltage across R will be reversed, i.e. the rectifier will conduct on negative half-cycles of the input. Thus, if the lower line (connected to B) is used as the reference or chassis line, a positive or negative output voltage with respect to the reference line may be obtained depending upon the rectifier connections.

The half-wave circuit represents the most economical rectifier arrangement and is used for supplying equipment demanding only small current and power.

**Full-wave Rectifiers**

In a full-wave rectifier circuit use is made of both half-cycles of the a.c. input voltage so that the output does not remain at zero voltage for a half-cycle period as it does in the half-wave circuit. One arrangement using a centre-tapped mains transformer T1 is shown in Fig.1.2 together with waveforms explaining the operation.

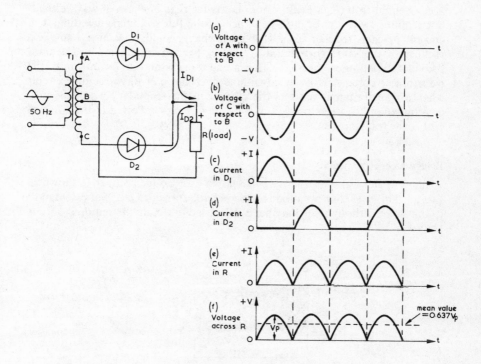

*Fig. 1.2 Full-wave rectifier circuit using centre tapped transformer.*

Two identical rectifiers D1 and D2 are now required supplying the common load represented by the resistor R, connected between the rectifier outputs and the transformer centre-tap. The secondary winding AC of T1 is a continuous winding with all its turns wound in the same direction but with an electrical centre-tap placed at point B. Thus, when the voltage at A is positive with respect to point B the voltage at C will be negative with respect to B and vice versa.

When A is positive with respect to B, a current $I_{D1}$ will flow through D1 and the load in the direction shown. During the half-cycle when A is positive with respect to B, the voltage at C is negative with respect to B and so no current will flow in D2 as it will be reverse biased. During the next half-cycle when A is negative with respect to B, the voltage at C will be positive with respect to B and current $I_{D2}$ will flow in D2 and the load in the direction shown. It will be seen that one of the diodes is conducting on each half-cycle of the input voltage and that current flows in the same direction through the load on every half-cycle, waveform (e). In consequence, there will be two half sinewaves of voltage developed across the load for each complete cycle of the input, waveform (f). The mean d.c. output voltage from the rectifier circuit is now equal to 0·637 × Peak Value, i.e. twice that of the half-wave circuit, but requires twice the peak-to-peak across the full secondary winding. The current in each rectifier is half of that which would flow in a half-wave rectifier of the same current rating. Since the current flowing in the total secondary winding is effectively alternating there is no d.c. saturation of the transformer core. This is an advantage over the half-wave rectifier circuit when a transformer is used as the current in the secondary would be uni-directional and cause saturation of the core (a factor reflected in the size and cost of the transformer).

### Bridge Rectifier Circuit

Another full-wave rectifier circuit known as a bridge rectifier is shown in Fig.1.3. This uses four p-n diode rectifiers but a transformer is not essential to its operation although one may have to be used to obtain the required output voltage.

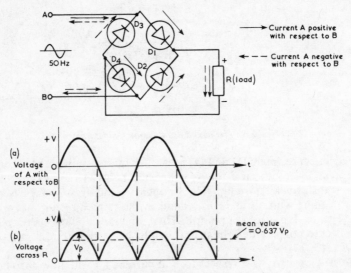

Fig. 1.3 Full-wave bridge rectifier circuit.

When the voltage at A is positive with respect to B, current will flow from A through D1, the load, D4 and back to B thus completing the circuit. Current cannot flow in D2 or D3 as they will be reverse biased. During the next half-cycle when A becomes negative with respect to the voltage at B, D2 and D3 become forward biased and D1 and D4 revert to the reverse bias state. Current thus flows from B, through D2, the load and D3 back to A thereby completing the circuit. Thus on each half-cycle of the input, two of the diodes are conducting and current is flowing in the same direction through the load. In consequence, there will be two half sinewaves of voltage developed across the load as in waveform (b). The mean d.c. voltage output of the circuit is 0·637 × Peak Value as for the previous full-wave circuit.

The bridge circuit has the advantage that when a transformer is required, the secondary winding need not be centre tapped. A disadvantage is that four rectifiers are needed and that neither of the d.c. output lines is common to the inputs A or B.

### Smoothing Circuits

The d.c. output of the basic rectifier circuits described is not smooth enough to act as a supply for electronic equipment. The pulsating output voltage would cause a varying performance of the electronic circuits to be fed and would result in the operation ceasing when the rectifier output dropped to a low level, i.e. when the voltage approaches zero. To produce an output voltage with only a small amount of fluctuation a smoothing circuit is required. The smoothing action can be carried out in two stages by the use of a reservoir capacitor and a filter circuit.

### Use of Reservoir Capacitor

The half-wave circuit will be considered first and is shown in Fig.1.4. A large value capacitor C1 (usually an electrolytic) has now been connected across the load R. The output voltage from the circuit without C1 has been shown dotted in the waveform.

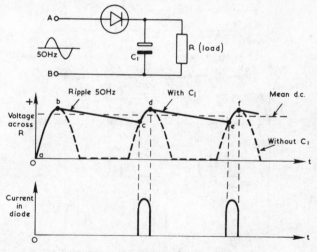

Fig. 1.4 Half-wave rectifier with reservoir capacitor.

In the first quarter-cycle period a-b when A is positive with respect to B and the diode conducts, current will flow into the output rapidly charging C1 via the low forward resistance of the diode. In consequence, the voltage across C1 will rise to the peak of the input voltage. At every instant the voltage across the load is the same as the capacitor voltage. Immediately following instant b the input voltage (which approximately follows the dotted line) commences to fall. Since the capacitor voltage does not fall as fast as the input, the rectifier becomes non-conducting. The load current is now supplied from C1 and the capacitor discharges during the period b-c. The rate at which the capacitor discharges depends upon the time constant of C1 and the load resistance. With a large value for C1 or a high value of load resistance (small current load) the slower will be the rate of voltage decay during b-c (also d-e). At instant c when the input voltage exceeds the capacitor voltage the rectifier becomes conductive and current flows in the rectifier recharging C1 to the maximum voltage. Immediately following instant d the rectifier again becomes reverse biased, current in the rectifier ceases and C1 discharges into the load.

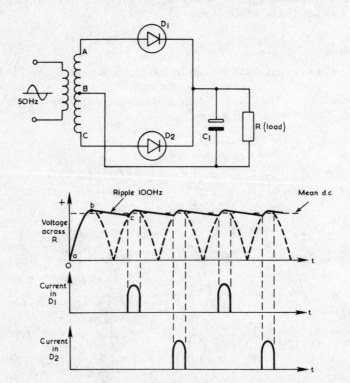

*Fig. 1.5 Full-wave rectifier with reservoir capacitor.*

It will thus be seen that the capacitor C1 acts as a store or reservoir of energy to supply the load current during the off period of the rectifier and for this reason is called a reservoir capacitor. The output voltage from the circuit is now as indicated by the solid line having the approximate shape of a saw-tooth waveform. The inclusion of C1 has now raised the mean d.c. level to almost the peak level of the input voltage and has reduced the amount of ripple in the output to around 10% of its previous value. The fundamental frequency of the ripple voltage is the same as the frequency of the input voltage, i.e. 50 Hz.

Current flows through the rectifier only during the brief periods corres-ponding to c-d and e-f, etc. (neglecting the first quarter period a-b) whereas current is supplied to the load continuously. Since the mean current passing through the rectifier must equal the mean current supplied to the load, the peak value of the current pulses in Fig.1.4 will be much greater than the mean current. For example, with a mean load current of, say, 80mA the peak current passing through the rectifier may be typically 650mA or higher. The peak current will increase as the value of C1 is made larger and hence there is a limit to the size of the reservoir capacitor in order that the peak current rating of the rectifier is not exceeded.

The effect of a reservoir capacitor on a full-wave rectifier is shown in Fig.1.5. An action similar to the half-wave circuit takes place but since the period b-c is approximately half that of the half-wave circuit the drop in voltage is about half that of the half-wave circuit (assuming the same CR product). Thus the ripple in the output is less which makes it easier to smooth. The fundamental frequency of the ripple is twice that of the supply frequency, i.e. 100 Hz, a factor contributing to the easier smoothing of the full-wave circuit. Since there are now two pulses of current for each cycle of the input, the peak current passing through each rectifier is half that of the half-wave rectifier circuit.

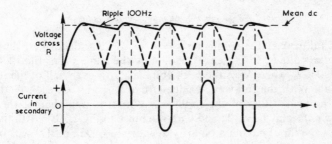

Fig. 1.6 Current in secondary winding of transformer feeding bridge rectifier.

In the bridge circuit the operation is identical but the current flowing in the secondary winding of the input transformer (a transformer will normally be used) consists of pulses acting in alternate directions, see Fig.1.6.

### Peak Inverse Voltage Rating

The maximum voltage that a rectifier will withstand in the reverse direction is known as the peak inverse voltage (p.i.v.), see Vol. 2. This maximum voltage depends upon the rectifier arrangement, see Fig.1.7. In the half-wave circuit of diagram (a) the reservoir capacitor charges up to the peak value $V_p$ of the input voltage during the conducting period of the rectifier with polarity as shown. During the non-conducting period the input voltage polarity will be as indicated resulting in a voltage between the anode and cathode of the rectifier of $2 \times V_p$ which must be within the p.i.v. rating of the chosen rectifier. For example, with a 240V mains input to the circuit, the maximum reverse voltage across the rectifier would be $2 \times 1 \cdot 414 \times 240V = 678V$. To avoid breakdown of the rectifier its p.i.v. rating must be above this figure, say, 1000V.

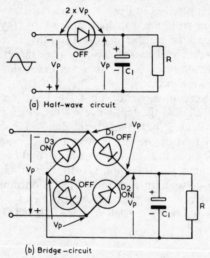

(a) Half-wave circuit

(b) Bridge –circuit

Fig. 1.7 Peak inverse voltage rating of rectifiers.

For the full-wave circuit utilising a centre-tapped transformer, the maximum reverse voltage across each rectifier is the same as in the half-wave circuit, i.e. $2 \times V_p$ where $V_p$ is the peak voltage across each half of the secondary winding. With the full-wave bridge circuit the p.i.v. of each rectifiier is equal to the peak value of the input $V_p$. Why it is half that of the other full-wave circuit may be explained with reference to Fig.1.7(b). With the peak value $V_p$ of the input having a polarity as indicated, D2 and D3 will be conducting and C1 will be charged to $V_p$ with polarity as shown. If D2 and D3 are thought of as closed switches it will be seen that the reverse voltage across the non-conducting rectifiers D1 and D4 is equal to $V_p$. Conversely, when the polarity of the input changes on the next half-cycle, D2 and D3 will be subjected to a maximum reverse voltage of $V_p$. This lower p.i.v. rating of the rectifiers used in a bridge circuit is an advantage over the full-wave circuit employing a centre-tapped transformer.

## Filter Circuits

The degree of smoothing produced by the reservoir capacitor is not usually adequate, so it is normally followed by a filter circuit (or a voltage stabiliser).

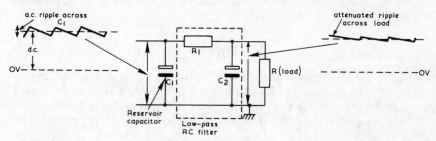

*Fig. 1.8 Resistance capacitance filter circuit.*

A common type of filter circuit found in equipment demanding small current is the resistance-capacitance filter shown in Fig.1.8. R1, C2 forms the filter (low pass), C1 the reservoir capacitor and R the load as previously discussed. The purpose of the filter is to reduce the ripple voltage present across C1 to an acceptable level across the load without attenuating the d.c. output voltage from the rectifier. Although the voltage at C1 is a fluctuating d.c. voltage it may be considered as consisting of a d.c. voltage (equal to the mean value of the waveform) and an a.c. component representing the ripple voltage. The ripple voltage is a complex waveform and as such it is composed of a fundamental sinewave component plus harmonics. Only the fundamental sinewave component will be considered because if adequate filtering of the fundamental occurs the harmonic components will be reduced to a very low level indeed. The d.c. and a.c. components will be dealt with separately.

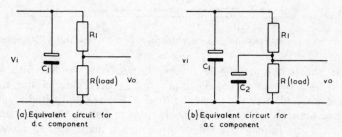

*Fig. 1.9 RC filter circuit drawn as a potential divider.*

Capacitor C2 will not pass d.c. thus the equivalent circuit for the d.c. component is as in Fig.1.9(a). It will be seen that R1 and the load form a potential divider for the d.c. voltage ($V_i$) across C1. Provided that the resistance of R1 is small compared with the load resistance the d.c. output voltage ($V_o$) will be practically equal to $V_i$ as required.

Considering now the a.c. component or ripple voltage $V_i$ across C1, capacitor C2 will pass this component and so is included in the equivalent circuit of diagram (b). The ease with which C2 will by-pass the ripple from the load depends upon the reactance of $C_2$ ($X_{c2}$), given by:

$$\frac{1}{2\pi fC_2}$$

where f is the frequency of the ripple. The smaller the reactance of C2 compared with the resistance of R1, the less will be the amount of ripple voltage developed across the load. As the value of C2 is increased, the magnitude of the output ripple is reduced but the maximum value for C2 is often limited by cost and space.

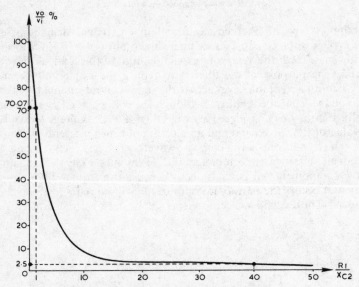

Fig. 1.10 Graph showing relationship between output ripple voltage and $R_1/X_{c2}$ ratio for RC filter.

Fig.1.10 shows how the ripple voltage output from the RC filter circuit (expressed as a percentage of the input ripple) decreases as the ratio of $R_1/X_{c2}$ is increased, assuming that the load resistance (R) is high compared with the reactance of C2. For given component values the filter circuit is more efficient in a full-wave rectifier output circuit than in a half-wave, since the ripple frequency is twice that of the half-wave rectifier.

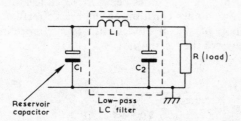

*Fig. 1.11 Inductive capacitance circuit.*

The low-pass filter may alternatively consist of a series inductor (L1) and a shunt capacitor (C2) as shown in Fig.1.11. The equivalent circuit for d.c. is as shown in Fig.1.12(a) where $R_{L1}$ is the d.c. resistance of the inductor. C2 has been omitted because a capacitor will not pass d.c. Provided $R_{L1}$ is small compared with R, most of the input voltage $V_i$ will appear across the load as is required.

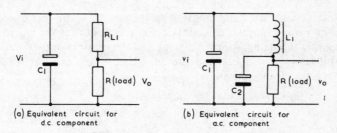

*Fig. 1.12 LC filter circuit drawn as a potential divider.*

For the a.c. component the equivalent circuit is as shown in Fig.1.12(b). Inductor L1 offers a high reactance to a.c. equal to $2\pi fL$, whereas C2 offers a low reactance (as for the RC filter). Since by choice of L1 value, the reactance of L1 is large compared with the reactance of C2, very little ripple voltage will appear across the load. With typical values for L1 and C2, the output ripple voltage will be 1/100th of the input ripple or less. LC filters are more commonly used in high voltage supplies (above 200V) as it is difficult to make the d.c. resistance of L1 low enough to prevent appreciable d.c. voltage drop across it. Although more efficient than an RC filter, the LC filter has the disadvantage that the inductor is costlier and takes up more space.

An example of a low voltage power supply is shown in Fig.1.13. Here a full-wave bridge rectifier is used fed from a step-down mains transformer T1. C1 is the reservoir capacitor followed by an RC filter R2, C2. The fuse F1 will rupture should the load current exceed 100mA or a fault occur in the power supply, e.g. C1 short-circuit. R1 is included to prevent the peak current rating of the rectifier diodes being exceeded at switch-on when C1 is uncharged. The four diodes constituting the bridge rectifier usually consists of an encapsulation of silicon diodes with four connections: two for the a.c. supply, and two for the d.c. output.

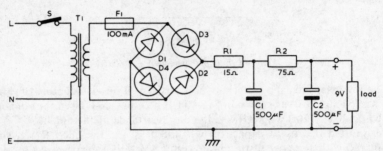

Fig. 1.13 Power supply incorporating bridge rectifier and smoothing circuit.

## Voltage Stabilisers

As explained in Vol. 1 *Electronic Systems*, voltage regulators or stabilisers are used on power supplies to maintain a constant output voltage in spite of (a) changes in the mains supply voltage; and (b) changes in the load current. The stabiliser is placed between the smoothing filter and the load so the input to it is d.c. as in Fig.1.14.

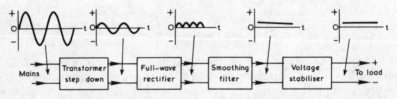

Fig. 1.14 Basic arrangement of low voltage stabilised supply.

A number of different circuit arrangements are possible for the stabiliser and a simple arrangement using a series resistor and shunt zener diode was described in Vol. 2 *Core Studies* (*Electronic Devices*), the circuit of which is shown in Fig.1.15.

A diode is chosen with a zener voltage corresponding to the required value of the stabilised output voltage ($V_z$). This voltage must be less than the unstabilised input voltage ($V_s$), the polarity of which must place the diode in the reverse bias state. The excess voltage between the input voltage and the output voltage ($V_s-V_z$) is dropped across the series resistor R. For good

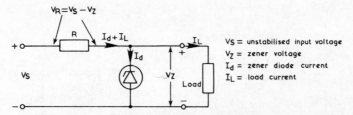

Fig. 1.15 Simple type of stabiliser using a zener diode and series resistor.

stabilisation the power rating of the diode should be such that it will carry a current ($I_d$) of about four times the expected load current ($I_L$). If $V_s$ varies, $I_d$ varies to alter the voltage drop across R to maintain a constant output voltage. For example, when $V_s$ increases, $I_d$ increases thereby increasing the voltage drop across R. On the other hand if $V_s$ decreases, $I_d$ decreases and there will be less voltage drop across R.

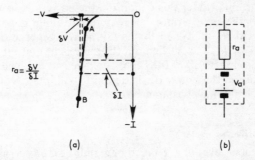

Fig. 1.16 Reverse characteristic of zener diode and its equivalent circuit.

In practice, the output voltage does not remain perfectly stable with changes in input voltage. The degree of stabilisation depends upon the slope resistance ($r_a$) of the diode, see Fig.1.16(a). A zener diode may be considered as a constant voltage source represented by a battery $V_A$ in series with its slope resistance $r_a$, see equivalent circuit of Fig.1.16(b). To changes of input voltage, the series resistor R and the slope resistance $r_a$ form a potential divider and provided $r_a$ is small compared with R there will be little change in the output voltage. The value of $r_a$ for a 9V zener diode when passing 10mA is typically 5 $\Omega$, whereas R may be, say, 800$\Omega$. It is often assumed that over the portion A–B of the reverse characteristic it is straight, but in practice it is more complex. For any particular diode the larger the current, the lower is the slope resistance.

The circuit will also stabilise against variations in the current drawn by the load. Should $I_L$ increase, the diode current would fall to try to maintain a constant voltage drop across R and hence a constant output voltage. On the other hand a fall in $I_L$ will cause a rise in the diode current, again trying to

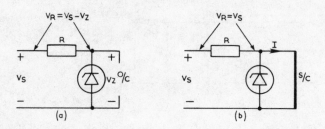

*Fig. 1.17 Zener diode stabiliser under o/c and s/c load conditions.*

maintain a constant voltage drop across R and hence a constant output voltage.

Under fault conditions with an o/c load as in Fig.1.17(a), the load current will fall to zero and the diode current will increase to $I_L + I_d$ where $I_L$ is the normal load current. Under this condition, the power dissipated in the diode will increase. If the power rating of the chosen diode is adequate then no adverse effect will result. When this is not so the zener diode may be destroyed. With a s/c load as in diagram (b) there is no output voltage and no current in the diode and all of the supply voltage is developed across R. Thus under this condition the power dissipated in R will increase, and if its power rating is insufficient it may be damaged.

### Transistor Stabilisers

An improvement over the zener diode shunt stabiliser can be obtained using a transistor as the stabilising element. The transistor may be connected in shunt with the load or in series with it. This has lead to designs based on two basic types of regulator.

### Transistor Shunt Stabiliser

A basic shunt stabiliser is shown in Fig.1.18. The transistor (n-p-n or p-n-p) is connected in shunt with the load with its base current supplied via a zener diode D1.

Variations in the supply voltage $V_s$ cause variations in TR1 current and in the voltage drop across R2. Since R2 is in series with the load, variations in its voltage drop will stabilise the output voltage $V_o$ because at all times $V_o = V_s - V_{R2}$. Suppose that $V_s$ increases (causing a larger current in D1 and hence a larger base current in TR1) the collector-to-emitter current of TR1 will increase and since this flows in R2 there will be a greater voltage drop across R2. If the increase in voltage drop in R2 is equal to the rise in $V_s$, the output voltage will remain constant.

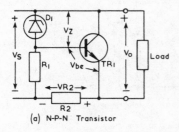
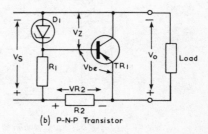

(a) N-P-N Transistor    (b) P-N-P Transistor

*Fig. 1.18 Transistor shunt stabiliser.*

It should be noted that the output voltage is equal to the zener voltage $V_z$ of D1 plus the base-emitter voltage drop $V_{be}$ of TR1. Since $V_{be}$ is normally small compared with $V_z$ the output voltage is approximately equal to $V_z$. Thus the only way of altering the magnitude of $V_o$ is by using a diode with a different zener voltage. The circuit will also stabilise against load current variations. Should, say, the load current increase it would increase the voltage drop across R2 since the load current flows in R2. As a result TR1 will conduct less (its $V_{be}$ will fall) and there will be less current in R2 due to TR1, thereby tending to maintain a constant voltage across R2 and hence a constant output voltage.

An advantage of the shunt stabiliser is that no damage is done to the transistor if the output terminals are short-circuited. This results in zero voltage across the transistor and the short-circuit current is limited by R2. Under open-circuit load conditions TR1 will conduct harder as there will be less voltage drop in R2 but the collector-emitter voltage will remain about the same. The design will normally cater for this condition and no harm will be done.

In a typical stabiliser of this type providing 9V d.c. output, variations in the mains supply to the power unit of between 190V to 250V will result in a change of output voltage of only 0·2V. Also, a load current variation from 0mA to 125mA produces an output voltage change of only 0·25V.

The circuits of Fig.1.18 operate by the change in TR1 base current when $V_s$ varies. As the base current is supplied via D1, changes in the current will alter the zener voltage and hence the output voltage. An improvement in the regulation may be achieved by adding another transistor (TR2) as in Fig.1.19.

The base current of TR1 is now supplied from TR2 emitter and the base current of TR2 will be smaller than its emitter current by the current gain of TR2. Thus if TR2 current gain is, say, 60, variations in D1 current are 60 times smaller than variations in TR1 base current, resulting in a more stable zener voltage. It should be noted that the output voltage is now $V_z + V_{be}(\text{TR2}) + V_{be}(\text{TR1})$.

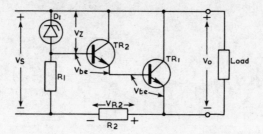

*Fig. 1.19 Shunt stabiliser using cascade transistors.*

## Transistor Series Stabiliser

With series type stabilisers it is arranged that any change in the unstabilised input appears across a series-connected stabilising transistor so that the output voltage across the load remains constant. An elementary circuit of this type is shown in Fig.1.20 for use with an n-p-n or p-n-p transistor.

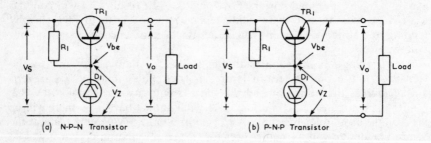

(a)  N-P-N  Transistor                         (b)  P-N-P  Transistor

*Fig. 1.20 Transistor series stabiliser (emitter-follower).*

TR1 is the regulating transistor which is supplied with a constant base voltage $V_z$ from the zener diode D1 fed with a suitable current via R1. As the load is connected in the emitter the arrangement forms a common collector or emitter-follower circuit, thus the emitter voltage follows the base voltage. Therefore, if the base voltage is held constant by D1, the emitter voltage and hence the output voltage will remain constant.

If, say, $V_s$ rises causing $V_o$ to try to rise, the $V_{be}$ of TR1 is reduced causing TR1 collector-to-emitter resistance to increase. As a result the increase in $V_s$ appears across the collector-to-emitter of the transistor thus restoring $V_o$ to an equilibrium value. It should be noted that the output voltage is equal to $V_z$ minus the $V_{be}$ of TR1. The only way of altering the value of the output voltage from this arrangement is by altering the zener voltage.

The circuit will also deal with variations in the load current. If, say, the load current increases the $V_{be}$ of TR1 will increase and the collector-to-emitter resistance of the transistor will be lowered. Thus less voltage is dropped across the series transistor which assists in keeping the output voltage constant. A disadvantage of the series transistor stabiliser is that if the load is short circuited, the full input voltage appears across the transistor and a large current flows. As this can lead to the destruction of the transistor some form of current limiting is desirable in series transistor stabilisers.

## Use of feedback

The regulation of the basic emitter-follower stabiliser can be improved by the use of amplified feedback and a basic circuit is given in Fig.1.21.

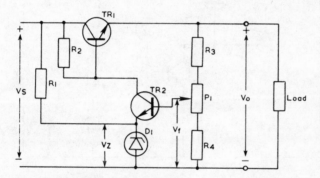

*Fig. 1.21 Transistor series stabiliser using feedback and amplifier.*

A potential divider comprising R3, P1 and R4 is now connected across the output and a fraction of the output voltage $V_f$ is fed back to the base of an amplifier TR2. Here $V_f$ is compared with a reference voltage $V_z$ developed across the zener diode D1 and applied to the emitter of TR2. If $V_o$ tries to depart from its steady value, $V_f$ will change causing a change in the difference voltage between $V_z$ and $V_f$. This change in voltage is amplified by TR2 and applied to the base of the series regulator TR1 in such a direction that the variation in volts drop across TR1 restores the output voltage to its nominal level. The zener diode is supplied with a suitable current via R1, and R2 serves as TR2 collector load as well as passing TR1 base current.

Suppose that $V_o$ tries to rise due to an increase in $V_s$ or a reduction in the load current. This will cause $V_f$ to rise resulting in an increase in the $V_{be}$ of TR2 (emitter voltage remains constant). TR2 will thus increase its conduction and there will be a larger volts drop across R2. In consequence, the base voltage of TR1 will be reduced causing TR1 to turn off thereby increasing its collector-emitter resistance. As a result there will be an increase in the voltage drop across TR1 thereby compensating for the rise in $V_s$ or the reduction in load current. The output voltage will be slightly greater than prior to the variation, but if the gain of TR2 is large the difference will be small. P1 provides a means of altering the output voltage. If P1 setting increases the base voltage of TR2 it will conduct harder increasing the steady voltage drop across R2. This will turn TR1 towards the off condition thereby increasing the collector-to-emitter resistance and lowering the output voltage. Moving P1 setting in the opposite direction will turn TR2 towards the off condition resulting in less voltage drop across R2 and TR1 turning harder on. In consequence there will be less voltage drop across TR1 collector-to-emitter and the output voltage will increase.

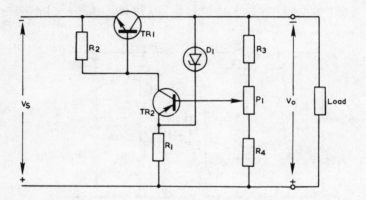

*Fig. 1.22 Another common series stabiliser with feedback and amplifier.*

Another commonly used series stabiliser is given in Fig.1.22 with TR1 connected in the common emitter configuration. The positions of D1 and R1 have been interchanged and are now fed from the stabilised side of TR1. An n-p-n transistor is still used for TR1 (but the supply has been reversed) whilst the amplifier TR2 is a p-n-p transistor. The principle of operation is similar to the previous circuit but the action is slightly different.

Suppose that $V_o$ tends to rise due to an increase in $V_s$ or a decrease in load current. Now, since D1 and R1 have been interchanged, the full rise in $V_o$ is felt across R1 and is applied to TR2 emitter (the voltage across D1 remains constant). Only a portion of the rise in $V_o$ is fed to TR2 base because of the action of the potential divider R3, P1 and R4. Thus because the emitter voltage rises more than the base voltage, the current in TR2 will be reduced. There is now less voltage drop across R2 causing TR1 to turn towards the off condition and for a larger voltage drop to occur across TR1 thereby restoring the output voltage to an equilibrium value. It will be seen that the action in TR2 is opposite to that of TR2 in the previous circuit, but TR1 operation remains the same except that the output is from the collector.

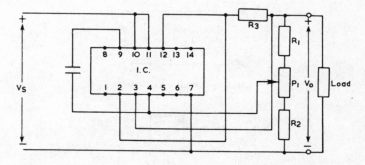

*Fig. 1.23 Monolithic integrated circuit series stabiliser.*

### I.C. Stabiliser

Complete voltage stabiliser circuits fabricated in integrated circuit form are now in common use, one arrangement being shown in Fig.1.23. The i.c. contains a series transistor stabiliser, a reference voltage source, an error amplifier and a current-limiting circuit.

A potential divider comprising R1, P1 and R2 is connected external to the i.c. with the feedback signal applied to one of the pins of the i.c. from P1 slider. Adjusting P1 alters the level of the stabilised output voltage. The inclusion of R3 in series with the load allows the load current to be sampled. If this rises above a predetermined value the current limit circuit comes into operation. One way of limiting the load current is to reduce the conduction of the series regulating transistor so that the output voltage is reduced once a certain current is exceeded. This protects the series regulating transistor from damage under short-circuit load conditions.

### Switched Mode Power Supply

The main disadvantage of the series regulating element of Fig.1.22 is that energy which is dissipated in the series element is lost, thereby lowering the efficiency, particularly so in a high-current power supply. Also, the heat generated in the series element raises the temperature inside the enclosure of the power supply or equipment, which is undesirable.

Instead of using a series element whose resistance is varied, what is required is a series element that is ideally either fully 'on' (zero resistance) or fully 'off' (open-circuit). When fully 'on' there is no power dissipated when passing current since there is no voltage drop and when fully 'off' there is no current flowing thus there is no power dissipated. The ideal form of series element is thus a kind of 'switch' which is rapidly closed and opened at regular intervals. By varying the 'closed' to 'open' time of the switch the mean power fed to the load can be varied.

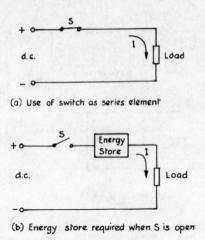

(a) Use of switch as series element

(b) Energy store required when S is open

Fig. 1.24 Idea of S.M.P.S.

Consider the circuit of Fig.1.24(a) where an electrical switch is used as the series element. If the switch is closed then a current I will flow in the load. However, if the switch is opened no current will flow in the load. Since continuous power is required by the load then some form of 'energy-storing' device is needed as in Fig.1.24(b) so that current may be supplied to the load when the switch is open.

An inductor may be used as an energy store and this is featured in the basic circuit of Fig.1.25(a). Additionally a diode D is required to provide a current

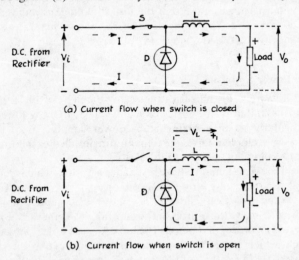

(a) Current flow when switch is closed

(b) Current flow when switch is open

Fig. 1.25 Basic circuit of switched mode stabiliser.

path when the switch is open. When the switch is closed, current I flows round the circuit and through the load in the direction shown producing an output voltage $V_o$ across the load. During this time the diode is reverse biased and therefore non-conducting. The current flowing in L causes a magnetic field to be set up and energy is stored in this field. If the switch is now opened as in Fig.1.25(b), the magnetic field around L collapses, inducing a voltage into the inductor ($V_L$). This voltage forward biases D causing the diode to conduct thereby allowing current to flow in the load in the same direction as previously. Thus when the switch is opened, the load uses the energy stored in L with D acting as an 'efficiency diode'.

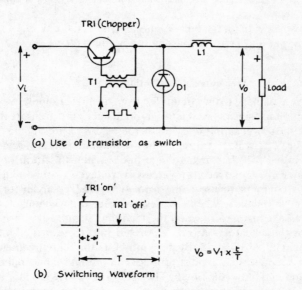

(a) Use of transistor as switch

(b) Switching Waveform

$$V_o = V_1 \times \frac{t}{T}$$

Fig. 1.26 Practical arrangement of switched mode stabiliser.

In a practical arrangement the electrical switch may be replaced by a transistor as in Fig.1.26(a) where the transistor is either fully 'on' or fully 'off'. To switch the transistor alternately 'on' and 'off' a pulse waveform may be supplied as indicated via a transformer T1. To regulate or stabilise the output voltage all that is necessary is to detect changes in the output voltage $V_o$ and correct it by altering the ratio t/T (the duty cycle) of the switching waveform shown in Fig.1.26(b), since the output voltage will depend upon how much current flows when the transistor is 'on'. The idea is illustrated by the basic block schematic of Fig.1.27. The d.c. output of the rectifier A is applied to the switching transistor B (referred to as the chopper). The chopper is driven with continuous pulses from E; a triggered monostable oscillator may be used to generate the pulses. The feedback amplifier D is used to detect changes in the output voltage and to adjust

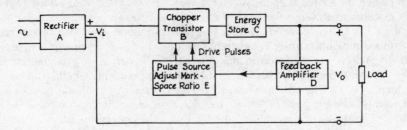

Fig. 1.27 Basic block diagram of switched mode power supply.

the mark–space ratio of the pulses generated in E, thereby varying the 'on' time of the chopper and hence stabilising the output voltage.

### Over-voltage and Over-current Protection

To prevent damage to the circuits fed from the power supply, for example i.c.s and other voltage-sensitive circuits, it is often necessary to ensure that the output voltage of the power supply does not increase above a preset limit for any reason whatsoever, e.g. voltage transients in the mains supply or accidental internal short-circuiting to other high-voltage lines. When this form of protection is incorporated it is referred to as **excessive voltage** or **over-voltage** protection. Also, it is common to use some form of **excessive current** or **over-current** protection, particularly when a series stabiliser or switching element is used, since excessive current drawn by the load may damage the series element.

Both of these features are incorporated into the switched-mode stabiliser block schematic of Fig.1.28. Block F monitors the output voltage and when it rises above a prescribed limit it produces an output that is applied to block G which 'trips' and blows the fuse F. The excessive voltage monitor may consist of a zener diode resistive network which produces an output when the zener

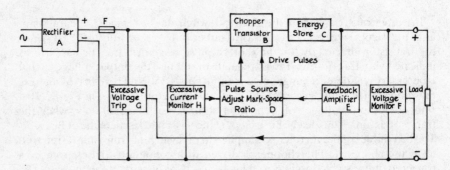

Fig. 1.28 Block schematic including over-voltage and over-current protection.

voltage is exceeded. The excessive voltage trip is commonly an s.c.r. which fires during over-voltage and acts as a 'crow bar' across the d.c. input to the stabiliser. The excessive current monitor H acts as a bypass for the switching current and when this exceeds prescribed limits the monitor removes the drive pulses to the chopper, thus quickly reducing the output voltage to zero under excessive current conditions.

### Voltage Doublers

In the rectifier circuits described earlier the d.c. output voltage is limited to the peak a.c. input to the rectifier arrangement. It is possible to provide d.c. outputs several times the peak a.c. input using voltage multiplier circuits. A circuit which provides a d.c. output twice that of the input voltage is called a voltage doubler and one type is shown in Fig.1.29.

During the half-cycles of transformer secondary voltage that make point A positive with respect to point B, D1 conducts causing a current $i_1$, to flow charging C1 to the peak secondary voltage $V_p$. On the other half-cycles when A is negative with respect to B, D2 conducts causing C2 to charge to $V_p$ by the current $i_2$. The d.c. voltage across the load is the sum of the voltages across C1 and C2, i.e. $2 \times V_p$. In effect the circuit is really two half-wave rectifier circuits in series resulting in twice the output voltage of a half-wave

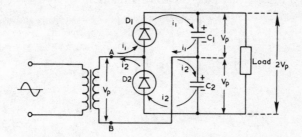

Fig. 1.29 Voltage doubler rectifier.

rectifier. C1 and C2 act as reservoir capacitors and if these have large values, and the load current is small, the output ripple will also be small. Since C1 and C2 are charged on alternate half-cycles the output ripple frequency will be twice that of the input frequency, i.e. 100 Hz with a 50 Hz mains supply.

An alternative doubler rectifier circuit, known as a cascade doubler is given in Fig.1.30. During half-cycles that make point A negative with respect to B, D2 conducts charging C1 with polarity shown to the peak voltage $V_p$. During these half-cycles D1 will be reverse biased and thus nonconducting. On the other half-cycles when A is positive with respect to B, no current will flow in D2 as it is reverse biased. However, the voltage across the secondary winding adds to the voltage across C1 causing D1 to conduct and for C2 to

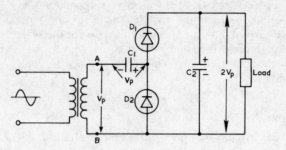

*Fig. 1.30 Cascade voltage doubler rectifier.*

charge. The total voltage applied to D1 is the peak secondary voltage $V_p$ plus the voltage across C1 (also $V_p$). Thus C2 charges up to $2 \times V_p$. During these half-cycles C1 is partly discharged but is recharged on the other half-cycles.

The cascade idea can be extended to produce tripling and quadrupling of the input voltage. Voltage doublers and triplers used in e.h.t. circuits in television receivers usually employ the cascade arrangement. Although they operate on similar lines to that shown in Fig.1.30 they are different in that the input to them is a pulse voltage waveform and not a sinewave of reversing polarity. An e.h.t. doubler with pulse input requires three diode-capacitor circuits and a tripler five diode-capacitor circuits.

## QUESTIONS ON CHAPTER ONE

Questions 1–7 refer to Fig.1.13 on page 12.

(1) If the current flowing in the load is 80mA, the d.c. voltage present between the junction of D2, D3 and chassis will be approximately:
   (a) 2V
   (b) 12V
   (c) 16V
   (d) 80V.

(2) A suitable wattage rating for R2 with 80mA load current would be:
   (a) 500mW
   (b) 250mW
   (c) 100mW
   (d) 50mW.

(3) If D1 were to go open circuit the outcome would be:
   (a) The voltage across the load would consist of half sinewaves only
   (b) The d.c. voltage across the load would be 4·5V
   (c) The voltage across the load would be greater than normal with an increase in the ripple voltage.
   (d) The voltage across the load would be slightly less than normal with an increase in the ripple voltage.

(4) With D1 open circuit there will be:
   (a) An increase in D2 peak current
   (b) An increase in the load current
   (c) An increase in D3 and D4 peak currents
   (d) A decrease in D3 peak current only.

(5) If R1 goes open circuit the effect will be:
   (a) The load current will fall to zero
   (b) The output voltage will fall to about 1V
   (c) D1 and D4 will conduct heavily
   (d) F1 will rupture.

(6) A short-circuit C1 will result in:
   (a) A large amount of ripple in the output
   (b) F1 rupturing
   (c) Smaller current in D1-D4
   (d) Load current increasing.

(7) An open-circuit C1 will result in:
   (a) No output voltage
   (b) F1 rupturing
   (c) Larger current in D1-D4
   (d) Low output voltage with an increase in ripple voltage.

Questions 8–11 refer to Fig.1.21 on page 17.

(8) A short circuit across the load will cause:
   (a) TR2 to conduct heavily
   (b) D1 to conduct heavily
   (c) Excessive output voltage
   (d) TR1 to conduct heavily.

(9) An open-circuit R3 will cause:
   (a) Zero output voltage
   (b) Smaller reduction in output voltage
   (c) Smaller current in TR1
   (d) Higher than normal output voltage.

(10) A short-circuited D1 will cause:
   (a) Low output voltage
   (b) Higher than normal output voltage
   (c) Zero output voltage
   (d) Small current in TR2.

(11) An open-circuited R2 will cause:
   (a) Zero output voltage
   (b) High current in TR1
   (c) High current in TR2
   (d) Higher than normal output voltage.

(12) The d.c. output voltage of a rectifier circuit may be made larger than the peak a.c. input to the rectifier by:
   (a) Using a mains transformer with a step-up ratio
   (b) Using a very large reservoir capacitor
   (c) Employing a fullwave rectifier
   (d) Using a voltage doubler.

# CHAPTER TWO

# DIFFERENTIATING AND INTEGRATING CIRCUITS

THE SHAPE AND AMPLITUDE of signal waveforms passing through the stages of an electronic circuit are particularly influenced by the presence of capacitance-resistance networks. In this chapter we shall consider how CR networks are used to achieve desired waveform shapes when fed with rectangular and sawtooth waves.

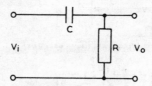

*Fig. 2.1 CR Network.*

Consider first the series CR network of Fig.2.1 when the rectangular pulses of Fig.2.2(a) are applied to the input. The shape and amplitude of the output voltage across R is determined by the relationship between the CR time and the pulse duration $T_p$. The CR time in seconds is given by the product of C (farads) and R (ohms). It will be assumed that the CR time is greater than the pulse duration. The response of the network will be explained with the aid of Fig.2.2(b) and the diagrams of Fig.2.3.

At instant $t_1$ when $V_i$ rises the voltages present in the network are as shown in Fig.2.2(a). As a capacitor cannot change its state of charge instantaneously, the full rise of the input voltage is initially developed across R, i.e. the capacitor passes the change of $V_i$ and thus the voltage across the capacitor is initially zero. With a voltage established across R a current i will flow as shown in circuit (b) causing C to charge. This occurs during the interval $t_1$—$t_2$. As C charges exponentially the voltage across R falls exponentially. If at sometime during this interval the voltage across C has risen to 0·1V the voltage across R will have fallen by 0·1V, i.e. from 10V to 9·9V. Since it was

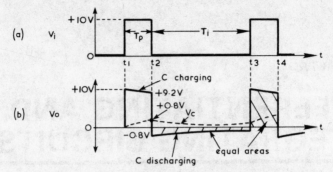

*Fig. 2.2 Response of CR network to rectangular pulse ($CR \gg T_p$).*

assumed that the CR time is greater than the pulse duration, the voltage across C will only have risen by a small amount, say, to 0·8V at the end of the pulse period. In consequence, the voltage across R will have fallen to 9·2V.

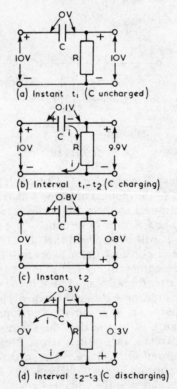

*Fig. 2.3 Diagrams showing voltages across R and C during pulse input.*

At instant $t_2$ the input voltage falls from $+10V$ to $0V$. This fall is passed through C to across R causing the voltage at the output to fall by 10V, from $+9.2V$ to $-0.8V$. This situation is illustrated in Fig.2.3(c) where the output voltage has reversed polarity to $-0.8V$ and the voltage present across C is still as was assumed at the end of the charge period (0.8V). The voltage across R now causes a current i to flow discharging C as shown in circuit (d). The capacitor discharges during the interval between pulses ($T_i$). Suppose that during the interval $t_2$–$t_3$ the voltage across C has fallen to 0.3V, the voltage across R will also have fallen to 0.3V. If the CR time is greater than the period $T_i$, then at the end of this period the voltage across C will not have fallen to zero but to some small voltage depending upon the period $T_i$ and the exact time constant.

It should be noted that at all times the algebraic sum of the voltages across C and R are equal to the input voltage $V_i$. When the next pulse arrives (instant $t_3$) and $V_i$ rises by 10V, the output voltage also rises by 10V causing C to charge once again repeating the operation. Note that the peak voltage at the output on the second pulse is slightly lower than for the first pulse due to the charge held by C at the end of the first discharge period. Succeeding peaks will gradually lower and then settle down to a constant level. At this stage the output waveform will balance itself about zero so that the pulse area above the zero datum line is equal to the pulse area below the datum line. Thus the d.c. component of the output wave form will be zero, which is only to be expected since the capacitor C cannot pass the d.c. component of the input waveform. The longer the time-constant, the more closely the output waveform will resemble the input waveform.

### Differentiation of Rectangular Wave

A differentiating circuit is one which produces an output that is directly proportional to the slope or rate of change of the input. The circuit of Fig.2.1 may act as a differentiator provided the CR time is very much shorter than the pulse duration. For practical purposes the CR time should be at least one-tenth of the pulse duration $T_p$.

The response of a differentiating circuit to rectangular pulses is shown in Fig.2.4. When the input voltage rises from 0V to $+10V$ at instant $t_1$, so does the voltage across R as previously explained. The voltage across R causes a current to flow charging C. Since the CR time is short compared with $T_p$, the capacitor charges rapidly and becomes fully charged during the pulse interval. As a result the voltage across R rapidly falls to zero. The output voltage remains at zero until instant $t_2$ when $V_i$ falls from $+10V$ to 0V at which point the voltage across R also falls by 10V, i.e. from 0V to $-10V$. The reversal of voltage across R causes a discharge current to flow and C rapidly discharges during the pulse interval $T_i$. As C discharges rapidly the voltage across R falls rapidly to zero.

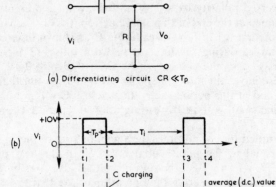

(a) Differentiating circuit CR ≪ Tp

Fig. 2.4 Response of differentiating circuit to rectangular input.

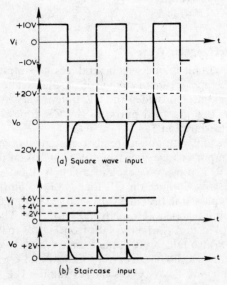

Fig. 2.5 Other examples of differentiation.

The differentiating circuit produces an output only when the input is changing its level; the output has a polarity that is linked to the rise and fall of the input. On the rising edge of the pulse a positive-going spike is produced but on the falling edge a negative-going spike is produced. There is no output when the input level is constant. The differentiating circuit is a high-pass filter passing only the rapid changes to the output. In the diagrams ideal pulses have been shown, i.e. having zero rise times or infinite slope. Instantaneous changes in voltage level cannot be obtained in practice.

Further examples of differentiation of pulse type waveforms are shown in Fig.2.5. In diagram (a) a square wave input is assumed swinging equally either side of its datum line. Output spikes only appear from the differentiator when the input level changes. When the input voltage falls from +10V to −10V (a fall of 20V) the output falls by 20V. Whereas on a rising edge of the input the voltage rises from −10V to +10V (an increase of 20V) and the output does likewise.

The response of a differentiator to a staircase waveform is shown in diagram (b). Note that each time the input voltage rises it changes its voltage level by 2V. Accordingly, the differentiator produces positive-going 2V spikes corresponding to these changes.

### Integration of Rectangular Wave

An integrating circuit is one which produces an output that is directly proportional to the area under the input waveshape. The idea of an integrator is shown in Fig.2.6.

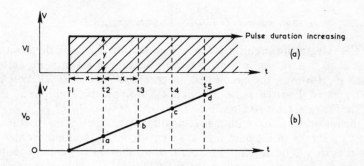

*Fig. 2.6 Basic idea of integrator.*

Consider that the input to the integrator is a rectangular pulse of increasing duration as shown in diagram (a). Prior to instant $t_1$, the pulse is of zero amplitude and the integrator output, diagram (b), is zero. At instant $t_2$, the pulse will have an area of $x \times y$ and it will be assumed that the integrator output is $a$ volts at this instant. If the pulse duration increases, then at instant $t_3$ the pulse area will be $2x \times y$, i.e. twice the area at $t_2$. The integrator output will now have increased to $b$ volts and will be twice that at $t_2$. At a later instant $t_4$ the pulse area will have increased to $3x \times y$ and the integrator output will have increased to $c$ volts and will be three times that at $t_2$ .... and so on. Thus the integrator produces an output that is proportional to the area under the pulse. The output voltage shows the relative contribution of succesive equal time increments to the total area under the pulse.

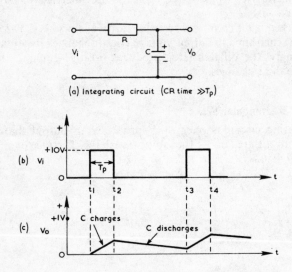

*Fig. 2.7 Response of integrating circuit to rectangular input.*

A CR integrating circuit may be formed by interchanging the positions of C and R to that shown in Fig.2.7(a) with the output taken from across the capacitor. However, the circuit will integrate only when the CR time is very much greater than the pulse duration $T_p$. For practical purposes, the CR time should be at least 10 times $T_p$.

If the integrator is fed with the rectangular wave of diagram (b), the output will be as shown in diagram (c). At instant $t_1$ when the input voltage rises, all of the rise will be developed across R and a current will commence to flow charging C with polarity shown. As the CR time is long compared with $T_p$, the charge acquired by C during the pulse period will be small and hence the voltage across C will also be small. The voltage rises fairly linearly during this period since only a small portion of the normal exponential charging curve is used.

At instant $t_2$ when the input falls to zero, the fall in voltage is developed across R causing a reversal of voltage across it and a current to flow in the opposite direction discharging C. If the CR time is also long compared with the interim period $(t_2-t_3)$, the voltage across C will fall only by a small amount. The fall in voltage will also be fairly linear since only a small portion of the normal exponential discharge curve is used.

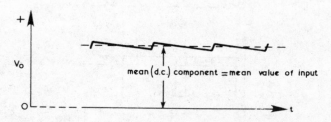

*Fig. 2.8 Voltage across C after a number of pulses.*

At the end of the discharge period, the voltage across C will not be zero, thus on the following pulse C will charge to a slightly higher voltage than on the first pulse. Gradually the peak voltage of $V_o$ will rise towards the mean value of the input with each succeeding pulse. After a number of pulses, the waveform across C will settle down to between constant levels as illustrated in Fig.2.8. The output waveform is a sawtooth having a d.c. component equal to the mean value of the input.

An integrator is a low-pass or smoothing filter, passing only the l.f components of the input pulse to the output.

### Differentiation and Integration of Sawtooth Wave

Differentiation of a sawtooth waveform is shown in Fig.2.9. Here the CR time must be short compared with the period $t_2-t_3$, in which case it will also be short compared with the period $t_1-t_2$. The sawtooth input consists of a rising ramp of relatively small slope followed by a falling ramp of much greater slope.

A differentiator produces an output that is proportional to the slope of the input as previously mentioned. Thus during the period $t_1-t_2$ the output is a constant voltage $v_1$, since the slope of the input is constant over this period. At instant $t_2$ where the input is at a peak, the slope is zero and therefore the output is zero. Between instants $t_2-t_3$ the output is a constant voltage $v_2$ but is of opposite polarity to $v_1$ since the slope is negative over this period. The

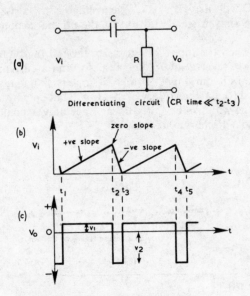

Fig. 2.9 Differentiation of sawtooth wave.

magnitude of $v_2$ is greater than $v_1$ since the slope is greater during the falling ramp period. Thus differentiation of a sawtooth wave produces a rectangular wave. A differentiating network can therefore be used to reverse the process of integration, since integration of a rectangular wave produces a sawtooth wave.

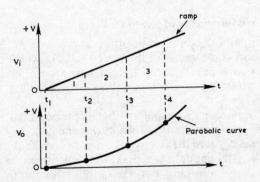

Fig. 2.10 Output of integrator with ramp input.

Before dealing with the integration of a sawtooth wave, consider the integration of the ramp voltage $V_i$ of Fig.2.10. Note that although the area under the ramp is increasing, the increase in area for successive time increments varies. Area 3 is greater than area 2 which is greater than area 1. Thus the output voltage $V_o$ will rise at a greater rate as the rise in the input voltage progresses, producing the parabolic curve shown.

When the input to the integrator is a sawtooth wave having rising and falling ramp periods as shown in Fig.2.11, the output consists of a series of parabolas as in diagram (c). During the rising ramp periods the parabolas ABC, EFG, etc., are produced, whilst during the falling ramp periods the inverted parabolas CDE, GHI, etc., are produced. These parabolas will have a d.c. component equal to the mean value of the input waveform after the circuit has reached its steady state condition.

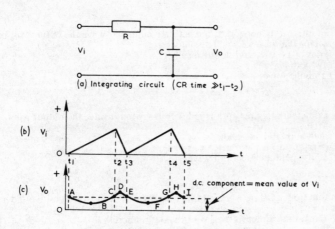

Fig. 2.11 Integration of sawtooth wave.

## QUESTIONS ON CHAPTER TWO

(1) A repetitive rectangular pulse waveform having a pulse duration of $50\mu s$ is fed to a CR network. The network will produce a differentiated output across R if its CR time is:
(a) 2 $\mu s$
(b) 50 $\mu s$
(c) 500 $\mu s$
(d) 5 ms.

(2) A repetitive rectangular waveform having a pulse duration of 2 ms fed to an integrating network. The CR time of the network would probably be:
(a) 20 $\mu s$
(b) 200 $\mu s$
(c) 2 ms
(d) 20 ms.

(3) When an integrating network is fed with a rectangular wave the output will be:
(a) A rectangular wave of small amplitude
(b) A series of positive and negative spikes
(c) A sawtooth wave
(d) A parabolic wave.

(4) When a differentiating network is fed with a sawtooth wave the output will be:
(a) A rectangular wave
(b) A small amplitude parabola
(c) A series of positive and negative spikes
(d) A small amplitude sawtooth wave.

(5) The output from a CR network fed with a square wave consists of positive and negative spikes when:
(a) An integrating circuit is used with the output across C
(b) A differentiating circuit is used with the output across R
(c) An integrating circuit is used with the output across R
(d) A differentiating circuit is used with the output across C.

# CHAPTER THREE

# OTHER DIODE CIRCUITS

THE SEMICONDUCTOR DIODE may be incorporated in a variety of electronic circuits to perform different operations on signal waveforms and some will be discussed in this chapter.

### D.C. Restoration

A common requirement in electronics is that of being able to alter the d.c. component of a waveform so as to set a particular level of the waveform to a specified d.c. voltage. This may be achieved using a d.c. restorer circuit. One such circuit which performs the operation of positive d.c. restoration is given in Fig.3.1.

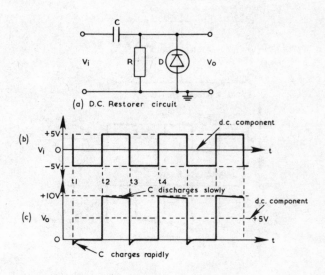

*Fig. 3.1 Positive D.C. restoration.*

The circuit of diagram (a) consists of a CR network with a diode D connected in shunt with the output. As we do not wish to alter the shape of the waveform in any way, the CR time must be very long compared with the periodic time of the input waveform. Consider a square wave input having zero d.c. component as in diagram (b).

When the input voltage goes negative at instant $t_1$, the change in voltage is initially developed across R. The direction of this voltage will cause D to be forward biased and the diode will conduct. When D conducts, C charges rapidly via the low forward resistance of the diode and the voltage across R quickly falls to zero. The output consists of a small negative going spike as shown in diagram (c). Between instants $t_1$–$t_2$ when the input is at a steady level the output is also steady at 0V since C is fully charged. At instant $t_2$ when the input suddenly rises by 10V (from $-5V$ to $+5V$), the full rise is developed across R. The polarity of the voltage across R will now reverse bias the diode and it will be non-conducting. However the voltage across R will allow C to commence to discharge during the period $t_2$–$t_3$. Since the CR time is very long compared with the periodic time, C will discharge only a little during this period and in consequence the voltage across R will fall only by a small amount. At instant $t_3$ when the input falls by 10V, the output also falls by 10V. In doing so the output will fall just below zero causing D to become forward biased once again. As D conducts, C rapidly charges replacing the charge lost during the discharge period. From instant $t_3$ onwards the operation is repeated.

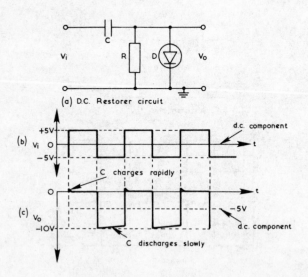

(a) D.C. Restorer circuit

(b) $V_i$

(c) $V_o$

Fig. 3.2 Negative D.C. restoration.

It will be seen that the most negative level of the input waveform has been shifted in the positive direction and has been restored to 0V. The d.c. component of the output waveform will therefore be +5V.

By reversing the diode connections, the circuit can be modified to perform negative d.c. restoration of the input signal as illustrated in Fig.3.2. Here the diode conducts on the positive rising edges of the input waveform causing C to rapidly charge but is reverse biased on the falling edges when C commences to slowly discharge. This time the input waveform has been shifted in the negative direction with its most positive level restored to 0V. The output waveform will therefore have a d.c. component of −5V.

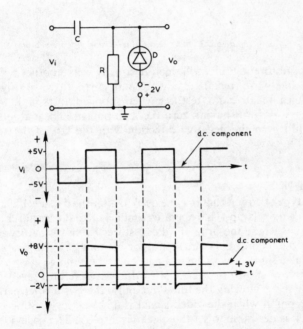

Fig. 3.3 Use of biased diode.

It may be required to restore a particular level of the input waveform to a voltage value other than 0V and this can be done by biasing the diode. The idea is shown in Fig.3.3 using a positive d.c. restorer circuit with a bias of 2V applied to the anode of the diode.

If the input waveform is the same as used previously, the diode will not commence to conduct until the voltage across R on the falling edges of the input is greater than 2V. Thus the most negative level of the input waveform will be restored to −2V instead of 0V. The output waveform will then have a d.c. component of $+3V$ (instead of $+5V$ when the diode is not biased). If the polarity of the bias voltage is reversed and the diode connections reversed then the circuit will operate as a negative d.c. restorer with the most positive level of the input waveform restored to $+2V$.

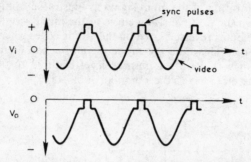

Fig. 3.4 Negative D.C. restoration of television video waveform.

These d.c. restorer circuits will operate equally well on other types of input waves, e.g. sine wave, rectangular wave and sawtooth wave. Fig.3.4 shows the action of a negative d.c. restorer when supplied at its input with a television video waveform having zero d.c. component. The output waveform has been shifted in the negative direction with the tips of the sync. pulses restored to 0V.

### D.C. Clamping

Another type of circuit that may be used to alter the d.c. level of a waveform is known as a d.c. clamp. This operates similarly to a d.c. restorer and is used in applications where the ordinary d.c. restorer is unsatisfactory. One application of a d.c. clamp is in the setting or clamping of the black level of a television video signal waveform to a particular d.c. voltage.

An example is shown in Fig.3.5 with the t.v. waveform given in diagram (a). Immediately following the line sync. pulse there is a brief period known as the back porch when the video signal is at black level. Suppose for simplicity that it is desirable to set the black level to 0V. The basic idea of a d.c. clamp is shown in diagram (b) with the t.v. waveform fed through the CR combination. If during the period corresponding to the back porch of the t.v. waveform the switch S is closed the capacitor will charge rapidly and point A will assume zero potential. Thus the black level of the waveform at the output will be set to 0V. When S is opened at the end of the back porch period, C will commence to discharge but if the CR time is very long compared with the period between line sync. pulses there will be little drift in the voltage corresponding to black level. In a practical circuit the mechanical switch S is replaced by an electronic switch; a diode may be used for this purpose as illustrated in diagram (c).

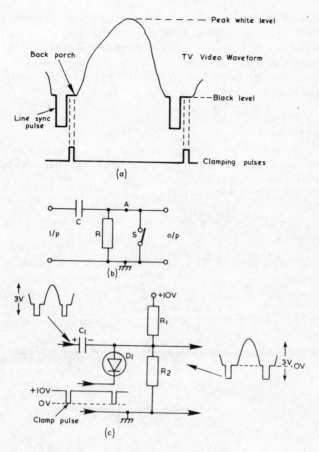

*Fig. 3.5 D.C. clamp circuit.*

To switch the diode on for a period corresponding to the back porch or a shorter duration, a clamping pulse is fed to the diode. This may be either a positive or negative pulse depending upon the circuit used, but negative clamp pulses are required here. Between clamping pulses D1 is reverse biased as its anode is returned to the potential divider R1, R2 connected across the 10V supply whilst its cathode is at +10V. When a clamping pulse arrives at D1 cathode, the cathode assumes zero potential and thus D1 conducts. This causes C1 to charge with polarity shown and for the voltage at the junction of R1, R2 to take up zero potential thereby clamping the black level to 0V. In between clamping pulses C1 discharges very slowly via R1 (a long CR time) thus the voltage corresponding to black level does not drift to any large extent. The next clamp pulse to arrive will clamp the voltage at the output back to 0V as the charge on C1 is restored.

A number of different circuit arrangements are possible using diodes but they all work on similar principles. In place of the diode a transistor may be used as the clamp. If the voltage level corresponding to the tip of the clamp pulse is made, say, $+2V$ the black level may be clamped to $+2V$ instead of 0V (assuming suitable values for R1 and R2). It is possible with a suitable circuit to clamp the black level to any chosen d.c. voltage.

### Limiting or Clipping

Another common requirement in electronic circuits is to be able to limit the amplitude of a waveform at a particular d.c. level. A limiter or clipper circuit may be used for this purpose.

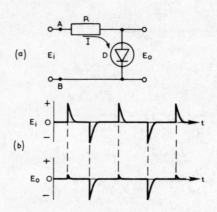

*Fig. 3.6 Diode clipper or limiter.*

As an example consider the circuit of Fig.3.6 (a) fed with a series of positive and negative-going spikes at its input as shown in diagram (b). If during the positive spikes terminal A is positive with respect to terminal B, the diode will conduct and a current will flow in R. If the resistance of R is large compared with the forward resistance of the diode then most of the spike voltage will be developed across R and little across the diode (the output voltage will be limited to the forward voltage drop of the diode, say, 0·8V for a silicon diode). During the negative spikes point A will be negative with respect to point B and the diode will be reverse biased. Thus the negative spikes will be passed to the output. Therefore the circuit has clipped off the positive-going spikes or limited them. If the connections of the diode are reversed, the circuit will clip off the negative-going spikes leaving only the positive ones at the output.

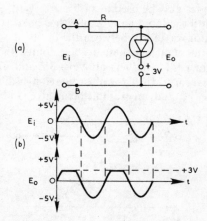

Fig. 3.7 Biased diode limiter.

If it is required to clip off a portion of a waveform, a biased diode limiter circuit may be used as shown in Fig.3.7 (a). Suppose that it is desirable to limit the positive half-cycles of the sinewave input to the circuit to, say, +3V. If a bias voltage of 3V is connected as shown, the diode will not conduct until point A is positive with respect to point B and the input voltage is greater than 3V. Thus the portion of the input waveform below the +3V level will be passed to the output. The voltage above this level will be dropped across R when the diode conducts.

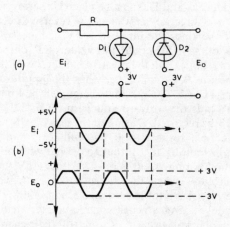

Fig. 3.8 Limiter with two biased diodes.

To obtain limiting on both half-cycles of the waveform to specific voltage levels, two biased diodes may be used as in Fig.3.8. With 3V biasing, neither diode will conduct until the input voltage of either polarity exceeds 3V. Thus the portion of the input waveform between the limits of −3V and +3V will be passed to the output. Outside these voltage limits, the excess voltage is dropped across R when D1 conducts above +3V and D2 conducts below −3V of the input waveform. In place of ordinary p–n diodes and bias supplies, zener diodes may be used, an example being given in Fig.3.9.

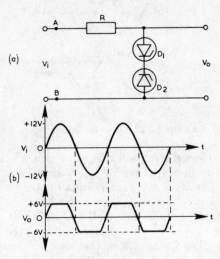

Fig. 3.9 Use of zener diodes for limiting.

The two zener diodes D1 and D2 must be placed in series since zener diodes conduct in the forward direction as well as the reverse direction. Suppose it is desirable to limit the sinewave input to ±6V. Diodes would be chosen with zener voltages of, say, 5·3V and the value of R chosen to provide a suitable current in the diodes. On the positive half-cycles of the input when A is positive with respect to B, D2 will zener when the input exceeds 6V and D1 will become forward biased. The output voltage will be limited to the zener voltage of D2 (5·3V) plus the forward voltage drop of D1, say, 0·7V, i.e. a limiting voltage of 6V as required. During the negative half-cycles of the input when A is negative with respect to B, D1 will zener and D2 will become forward biased, again limiting the output to 6V but with opposing polarity. Above and below the limiting voltages of +6V and −6V the excess voltage will be dropped across R when the diodes are conductive. As the zener characteristic is not very sharp for low voltage zener diodes, the circuit works best with zener diodes having a zener voltage greater than about 5V. To provide limiting on one half-cycle only, one of the zener diodes may be replaced by an ordinary p–n diode.

### Diode Protection

Another common use for diodes is in protection circuits where they are employed to protect other components from damage due to over-voltage or voltage of incorrect polarity.

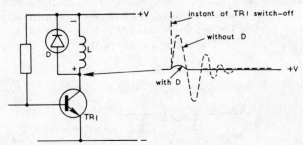

(a) Diode used to protect transistor against induced voltage of L

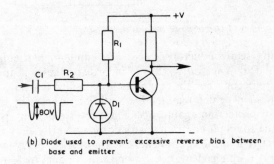

(b) Diode used to prevent excessive reverse bias between base and emitter

*Fig. 3.10 Diode protection circuits.*

Two examples of over-voltage protection are given in Fig.3.10. When an inductor L is used as a load for a transistor as in diagram (a), excessive voltage may be applied between collector and emitter when the current in the transistor is cut off. Without the diode a large induced voltage (with polarity shown) may appear across the inductor when the current is cut off or rapidly reduced. This may cause the collector voltage to swing well above the $+V$ supply line as shown causing the maximum collector-to-emitter voltage rating to be exceeded and for the transistor to be destroyed. When a diode is fitted with connections as shown, any attempt for the collector voltage to rise above the supply line potential is soon arrested as it will bring the diode into forward conduction. The conduction of the diode quickly dissipates the energy stored in the inductor and prevents the rise of collector voltage. The inductance L may be the primary winding of a transformer, the operating coil of a relay or a choke. For a p–n–p transistor and a negative supply line the connections of the diode would be reversed.

In diagram (b) the diode is used to prevent the maximum reverse voltage across the base-emitter junction being exceeded when the base is fed with a large amplitude pulse to cut off the transistor. Forward bias for the transistor is provided by R1 and the base-emitter forward voltage drop places D1 in the reverse bias state so it has no effect. However, when the negative pulse is applied through C1 and R2 to cut off the transistor, D1 conducts and limits the maximum reverse voltage to its forward voltage drop of, say, 0·8V thus protecting the transistor. The excess voltage of the pulse is developed across R2.

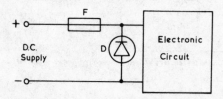

Fig. 3.11 Use of diode for polarity protection.

Circuits employing semiconductor devices (including in particular integrated circuits) may be damaged if voltage of incorrect polarity is applied to them. This is likely to occur when the equipment is supplied from an external battery the connections of which may inadvertently be reversed by the user. One arrangement for protection against accidental reverse polarity is shown in Fig.3.11. When the supply is connected with correct polarity as shown, the diode is non conductive and the equipment works normally. If, however, the polarity of the supply is changed over, the diode will conduct thereby shorting out the supply and blowing the fuse F.

## Diode Demodulator

The purpose of the demodulator or detector in a receiver is to recover the original signal information impressed on the carrier at the transmitter. In the case of a broadcast radio transmission, the signal information consists of music and speech (audio) and either amplitude or frequency modulation of the carrier wave may be used. Here, we shall only be concerned with demodulation of a.m. waves as in a LW/MW radio receiver. The demodulator circuit used in this type of receiver is similar to that used for rectification in a power supply and operates almost identically. A basic circuit with waveforms is given in Fig.3.12.

The circuit is shown in diagram (a). The diode D1 (point contact type) is used as the demodulator and is fed at the input with the a.m. wave shown in diagram (b) from across the tuned circuit L1, C1 which is commonly the secondary of the final i.f. transformer. If, on the positive half-cycles of the modulated carrier input point A is positive with respect to point B, D1 will conduct and a current will flow in the diode and R1. When point A is negative with respect to point B on the negative half-cycles of the input, the diode will be reverse biased and there will be no current flow.

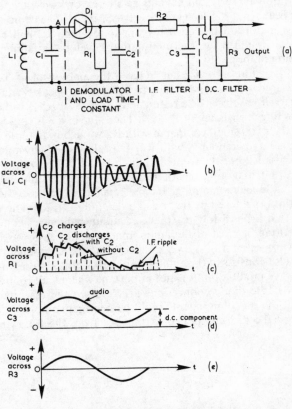

*Fig. 3.12 A.M. diode demodulator.*

Assuming that C2 is not fitted, the voltage across R1 would be as shown dotted in diagram (c) and consist of positive half-cycles only. It will thus be seen that the first operation performed on the modulated carrier is one of rectification (half-wave). To obtain maximum voltage across R1 which forms the d.c. load for D1, the value of R1 should be large compared with the forward resistance of the diode. The output of positive half-cylces contains: an i.f. component (unwanted); a d.c. component (unwanted); and the audio information (wanted). Without C2 the unwanted i.f. component would be of large amplitude and the wanted audio component of small amplitude. To improve the circuit efficiency C2 is added. The waveform across R1 is then as shown by the solid line in diagram (c). When the diode conducts on the positive half-cycles, C2 charges to the peak value of the input via the diode. During the negative half-cycles C2 discharges via R1. If the CR time of C2, R1 is long compared with the period for a half-cycle of the i.f. carrier, the voltage will fall only by a small amount during the non-conduction of the diode. It will thus be seen that C2 acts as a reservoir capacitor, charging on the positive peaks and discharging in between them. The choice of CR time for C2, R1 is important in that if it is made too long C2 will be unable to discharge fast enough when the modulation envelope is falling and would cause distortion of the audio.

Note that the addition of C2 has reduced the amount of i.f. ripple across R1 and increased the wanted audio to the peak-to-peak variation of the modulated carrier envelope. The residual i.f. ripple across R1 is then removed by a low-pass filter consisting of R2, C3. The value of C3 is chosen so that its reactance at the i.f. is small compared with R2 value but at audio frequencies its reactance is large compared with R2. This allows the unwanted i.f. to be developed across R2 but the wanted audio across C3. The voltage across C3 is then as shown in diagram (d) with the i.f. ripple removed and only the audio and a d.c. component remaining. It now remains to remove the d.c. component and this is achieved by the use of a d.c. filter C4, R3. The value of C4 is chosen so that it has a low reactance at audio frequencies compared with R3 value resulting in most of the audio appearing across R3 as is required. Since a capacitor will not pass d.c., C4 blocks the d.c. component so that the output across R3 consists of the audio component only as in diagram (e).

With sinewave modulation it is not important which way round the diode is connected. If the diode connections are reversed the essential difference would be that the d.c. component would be of negative polarity (instead of positive). Since the d.c. component is blocked from the following stage its

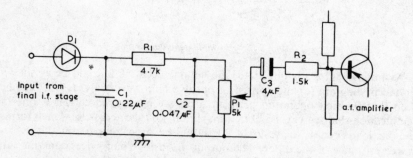

*Fig. 3.13 Typical practical A.M. demodulator circuit.*

polarity is of no consequence. However, the d.c. component has an amplitude proportional to signal strength and in some receivers it is used for a.g.c. purposes. In this use the polarity of the diode connections would be important.

In practice the demodulator circuit is usually arranged as in Fig.3.13 which shows typical component values for an a.m. transistor radio receiver. D1 is the demodulator diode with the series R1 and P1 acting as the d.c. load. C1 is the detector reservoir capacitor and R1, C2 form the i.f. filter. The d.c. component is blocked from the following stage (where it would upset the a.f. amplifier bias) by C3. The series resistor R2 'stands off' the low input resistance of the a.f. amplifying transistor from the demodulator circuit to prevent distortion of the audio signal. As well as forming part of the d.c. load for the demodulator, P1 also acts as the volume control.

## QUESTIONS ON CHAPTER THREE

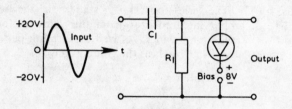

*Fig. 3.14 D.C. restorer circuit.*

(1)  The output from the circuit of Fig.3.14 above will be:

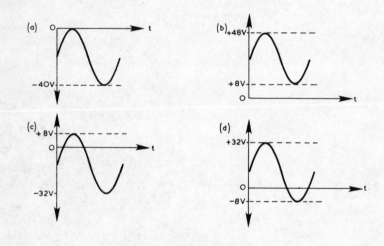

*Fig. 3.15.*

(2)  If the diode in Fig.3.14 above goes open circuit the effect will be:
    (a) No output waveform
    (b) Output almost identical to input
    (c) No charge path for C1
    (d) The positive peaks will be restored to 0V.

(3) To set the black level of a television waveform to a particular voltage level, which of the following would be used:
(a) A d.c. clamp
(b) A diode limiter
(c) A clipping circuit
(d) A vision demodulator.

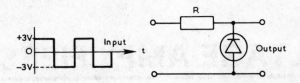

Fig. 3.16 Limiter Circuit.

(4) Neglecting the d.c. voltage drop across the diode, the d.c. component present in the output of Fig.3.16 above will be:
(a) +3V
(b) 0V
(c) −3V
(d) +1·5V.

(5) If the diode connections in Fig.3.16 are reversed, the d.c. component present in the output will be:
(a) −1·5V
(b) −3V
(c) +1·5V
(d) +3V.

(6) A diode may be fitted across an inductor in the collector circuit of a transistor to:
(a) Stop the transistor 'bottoming'
(b) Protect the transistor from over-voltage
(c) Reduce the output capacitance
(d) Stabilise the collector current.

(7) A low-pass filter is used in an a.m. demodulator circuit to:
(a) Remove the audio signal
(b) Block the d.c. component
(c) Reduce the i.f. ripple
(d) Remove hum voltages.

# CHAPTER FOUR

# VOLTAGE AMPLIFIERS

THE BASIC IDEA of a resistance loaded voltage amplifier employing bipolar and unipolar transistors was described in Volume 2 of this series. Methods of biasing the transistor, determination of voltage gain and stabilisation of the working point were discussed. It is now necessary to consider some aspects of multistage amplifiers, in particular their gain, method of coupling and frequency response.

### Cascade Amplifiers

If a voltage gain is required that is more than can be obtained from a single transistor amplifier, two or more stages may be connected together in cascade, i.e. the output from one stage feeding the input of the following stage. The idea is illustrated in Fig.4.1 which shows three separate amplifier stages connected in cascade.

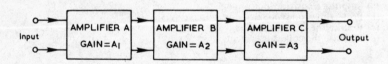

Fig. 4.1 Use of amplifiers in cascade to increase gain.

The overall voltage gain ($A_v$) of a cascade multistage amplifier is equal to the product of the individual voltage gains. For Fig.4.1 $A_v = A_1 \times A_2 \times A_3$. If, for example, each stage has a voltage gain of 40, then for two cascaded stages $A_v = 40 \times 40 = 1600$ and for three cascaded stages $A_v = 40 \times 40 \times 40 = 64,000$. The figures given for the voltage gain of each amplifying stage should be assumed to be the voltage gain measured from the input of one stage to the input of the following stage and not the voltage gain of an isolated amplifying stage. This is important as the voltage gain of any of the individual stages may be reduced when they are connected to the following stage.

Instead of expressing the individual stage gain or overall gain of an amplifier as a pure number, as in the above example, it is frequently given in decibels. This unit which may be used to express gain or attenuation will now be considered.

### Use of Decibels

Suppose that we have an amplifier as in Fig.4.2 fed with a voltage $V_i$ at its input. If $R_{IN}$ is the input resistance a current $I_i$ will flow when $V_i$ is applied. The input power $P_i$ is given by $V_i \times I_i$. Assume that the amplifier delivers an output voltage $V_o$ which is applied to the load $R_{LOAD}$ causing an output current $I_o$ to flow. The output power $P_o$ is given by $V_o \times I_o$.

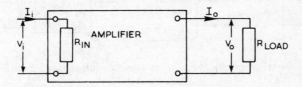

Fig. 4.2 Amplifier supplied with input power ($V_i \times I_i$) and developing output power ($V_o \times I_o$).

The power gain may be expressed as:

$$\frac{P_o}{P_i}$$

which will give a pure number. Alternatively, the power gain may be expressed as:

$$\log \frac{P_o}{P_i}$$

i.e. determine the ratio of:

$$\frac{P_o}{P_i}$$

and then find the logarithm (using tables or a pocket calculator). The unit of this ratio is called a bel; since this a rather large unit we use a decibel which is $\frac{1}{10}$ of a bel.

The power gain is then $10 \log \dfrac{P_o}{P_i}$ decibels (dB)

For example, if $\dfrac{P_o}{P_i}$ is equal to 32, the power gain in decibels will be:

$$10 \log 32 \text{dB}$$
$$= 10 \times 1 \cdot 505 \text{dB}$$
$$\simeq 15 \text{dB}.$$

The use of the decibel unit has a number of advantages.

(1) A convenient number is obtained when the power ratio is large
e.g. if the power ratio is 1,000,000, then in decibels it will be:

$$10 \log 1,000,000 \text{dB} = 60 \text{dB}.$$

(2) If the gains of individual amplifier stages are expressed in decibels, the overall gain of a cascaded amplifier is the sum of the gains of individual stages (adding logarithms of numbers is the same as multiplying the numbers together).

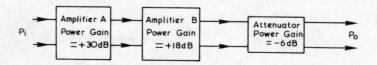

Fig. 4.3 Overall gain in decibels obtained by adding the gains of individual stages.

An example is given in Fig.4.3 which shows two amplifiers in cascade. If the power gain in decibels of amplifier A is $+30$dB and that of amplifier B is $=18$dB, the overall power gain of the two amplifiers is $30+18$dB$=48$dB. Note that if the output power exceeds the input power, i.e. a true power gain, the logarithm will give a positive result ($+$dB). Now suppose that the power output of amplifier B is fed to some form of attenuator (for example a cable to transfer the signal power) which introduces a power loss, i.e. the power output from the attenuator is less than the power fed into it. In this case the power gain in decibels will produce a negative result ($-$dB). Assume that the attenuator introduces a power gain of $-6$dB (a loss). The overall power gain from the input of amplifier A to the output of the attenuator.

$$\text{i.e. } \frac{P_o}{P_i}$$

expressed in decibels will be $+30$dB$+18$dB$-6$dB$=42$dB. Note that the adding is algebraic, allowing for negative signs.

(3) The human hearing sense grades loudness in a way which approximates to the decibel scale. The decibel is a convenient size as it is about the smallest change in sound intensity that the ear can detect.

Although the decibel is based on power rations, it may be used to express voltage or current gain. If the resistances are the same at the input and the output of an amplifier, the power is proportional to:

$$V^2 \left(\text{power} = \frac{V^2}{R}\right) \text{ or to } I^2 \left(\text{power} = I^2 R\right)$$

Thus we can express voltage gain as

$$10 \log \frac{V_o^2}{V_i^2} \, dB$$

$$= 20 \log \frac{V_o}{V_i} \, dB$$

and current gain as

$$10 \log \frac{I_o^2}{I_i^2} \, dB$$

$$= 20 \log \frac{I_o}{I_i} \, dB$$

| A | B | C |
|---|---|---|
| Output/Input ratio | Power (dB) | Voltage & Current (dB) |
| $10^6$ : 1 | +60 | +120 |
| $10^5$ : 1 | +50 | +100 |
| $10^4$ : 1 | +40 | +80 |
| 1024 : 1 | +30 | +60 |
| 512 : 1 | +27 | +54 |
| 256 : 1 | +24 | +48 |
| 128 : 1 | +21 | +42 |
| 64 : 1 | +18 | +36 |
| 32 : 1 | +15 | +30 |
| 16 : 1 | +12 | +24 |
| 8 : 1 | +9 | +18 |
| 4 : 1 | +6 | +12 |
| 2 : 1 | +3 | +6 |
| 1.414 : 1 | +1.5 | +3 |
| 1 : 1 | O | O |
| 1 : 1.414 | −1.5 | −3 |
| 1 : 2 | −3 | −6 |
| 1 : 4 | −6 | −12 |
| 1 : 8 | −9 | −18 |
| 1 : 16 | −12 | −24 |
| 1 : 32 | −15 | −30 |
| 1 : 64 | −18 | −36 |
| 1 : 128 | −21 | −42 |
| 1 : 256 | −24 | −48 |
| 1 : 512 | −27 | −54 |
| 1 : 1024 | −30 | −60 |
| 1 : $10^4$ | −40 | −80 |
| 1 : $10^5$ | −50 | −100 |
| 1 : $10^6$ | −60 | −120 |

*Fig. 4.4 Table of decibel values.*

As we are more usually concerned with measuring voltage or current, these expressions are commonly used especially the one for voltage gain. A table of decibel values for power, voltage and current gain is given in Fig.4.4. In column A we have various ratios of output/input, for ratios greater than unity (true gain) and less than unity (a loss or attenuation). Column B shows the decibel values for power ratios and column C shows the decibel values for voltage and current ratios. Note that for power ratios a change of 3dB represents a doubling or halving of the power and for voltage or current ratios a change of 6dB represents a doubling or halving of voltage or current. For example, if a signal of 10mV is applied at the input of a voltage amplifier having a voltage gain of +6dB the output voltage will be 20mV; with a gain of +12dB the output voltage will be 40mV; and with a gain of +18dB the output voltage will be 80mV . . . and so on giving a doubling of the input voltage for every 6dB increase in voltage gain.

### Cascade Amplifier with R–C Coupling

One method of coupling amplifier stages together in cascade is to use Resistance–Capacitance (R–C) coupling and an example is given in Fig. 4.5. Here two common emitter n-p-n transistor amplifier stages with identical component values are coupled together via R3 and C3 which form the R–C coupling. This is sometimes called capacitor coupling, since coupling is essentially via C3. The use of capacitor coupling allows the collector of one transistor to be coupled to the base of the following transistor as regards a.c., without the d.c. voltage on the collector of one stage upsetting the d.c. conditions on the base of the next stage, since a capacitor will not pass d.c. Because of the presence of C3 or any other capacitor (such as C1) in series with the signal path, the amplifier may be used only with a.c. signals.

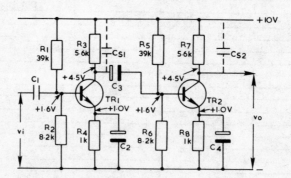

*Fig. 4.5 R-C coupled transistor amplifier.*

Both stages are biased to class A using the potential divider and emitter resistor method described in Volume 2. Silicon type transistors are assumed with base-emitter voltage drops of 0·6V. The input signal $v_i$ is applied via C1 (which blocks any d.c. component of the signal source) to the base of TR1. C1 must have a low reactance at all signal frequencies to be passed by the amplifier. An amplified signal voltage is developed across R3 and this is applied to TR2 base via the coupling capacitor C3. The reactance of C3 must be low compared with the input resistance of TR2 input circuit which is composed of R5, R6 and TR2 base-emitter resistance all in parallel with one another. If the lowest frequency to be passed by the amplifier is, say, 160Hz and the effective input resistance of TR2 input circuit is 1·2kΩ, then the reactance of C3 should be at least $\frac{1}{10}$ of the input resistance, i.e. 120Ω.

Now the reactance of a capacitor is given by $X_c = \dfrac{1}{2\pi f\, C}$

$$\therefore C = \frac{1}{2\pi f\, X_c}$$

$$C = \frac{10^6}{6\cdot284 \times 160 \times 120}\,\mu F$$

$$C \simeq 8\mu F.$$

Such a large value requires the use of an electrolytic type capacitor and due regard must be given to its polarity when connecting it into circuit (note polarity of C3). Because C3 is an effective short-circuit to signals, the input resistance of TR2 base circuit is effectively in parallel with R3 which will lower the effective load of TR1 stage and reduce its gain. Allowance must be made for this in determining the overall voltage gain required when amplifiers are coupled together.

If the lowest frequency to be amplified is, say, 100kHz then the value of C3 can be made accordingly smaller e.g. $0\cdot02\mu F$ and an electrolytic type capacitor would not be required.

The signal voltage applied to TR2 base is now amplified by TR2 stage which develops an output signal voltage across R7. This output signal voltage $V_o$ may be taken as shown from between TR2 collector and the negative supply line or from between TR2 collector and the positive supply line. From the signal point of view it makes no difference since the positive and negative supply lines are normally at the same potential as regards a.c. The overall voltage gain of the two stages is given by the ratio of

$$\frac{V_o}{V_i}\ \text{or in decibels by } 20 \log \frac{V_o}{V_i}$$

(Strictly speaking input and output resistance should be the same. However, gain is often expressed in decibels when input and output resistance are not the same because it is convenient.)

The emitter decoupling capacitors C2 and C4 are used to prevent the gain of their respective stages being reduced by a.c. negative feedback, which will be dealt with later. For the moment it is sufficient to say that the reactance of these capacitors must be low compared with the resistors they are de-coupling at the lowest signal frequency. Generally, their reactance is made about $\frac{1}{10}$ of the emitter resistors. With $1k\Omega$ emitter resistors this would indicate the use of electrolytic capacitors because of the large capacitance required for signal frequencies below about 2kHz.

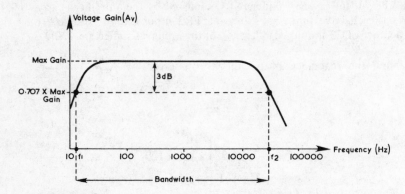

Fig. 4.6 Gain frequency response of R-C coupled amplifier.

The gain-frequency response of an R–C coupled amplifier is shown in Fig.4.6. At high frequencies the gain falls off due to the various stray capacitances present. These capacitances $Cs_1$ and $Cs_2$ are made up of the stray capacitance of the circuit and the transistor capacitances. At high frequencies these capacitances have a reactance which is low enough to appreciably shunt the collector load resistors causing a reduction of the voltage gain of the individual stages. At low frequencies the gain falls off due to the rising reactance of the coupling capacitor(s) which causes signal voltage to be dropped across them. Additionally, the rising reactance of the emitter decoupling capacitors introduces some negative feedback at low frequencies and hence some reduction in voltage gain.

The useful bandwidth of such an amplifier is the frequency-space between the frequency limits where the gain has fallen to 0·707 of its mid-band or steady reference gain level, i.e. in Fig.4.6 the bandwidth lies from $f_1$ to $f_2$. It will be noted from column C of Fig.4.6 that the ratio

$$\frac{1}{1·414} \, (0·707)$$

represents $-3dB$ voltage attenuation, thus at $f_1$ and $f_2$ the voltage gain will be 3dB down. Now, since power is proportional to $(voltage)^2$, the power will be $(0.707)^2$ or $0.5$ of its mid-band level; thus $f_1$ and $f_2$ are also referred to as the 'half-power' points.

When a comparatively wide bandwidth is required as in an audio amplifier, $f_1$ and $f_2$ may be, say, 30 Hz and 16 kHz respectively. If the amplifier is to be used to amplify a single frequency, e.g. a single tone test signal or the output from a transducer in electronic control equipment a much narrower band-width than that indicated could be tolerated.

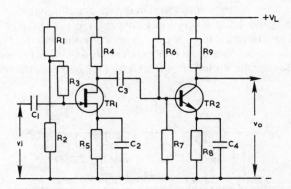

Fig. 4.7 Use of F.E.T. as first stage in R-C coupled amplifier.

Field effect transistors may be cascaded as for bipolar devices and use capacitor coupling between stages. However, since the voltage gain of an f.e.t. amplifier is appreciably less than a bipolar transistor amplifier, cascaded f.e.t. stages are not commonly used when high gain is required. An f.e.t. amplifier may be used with advantage as the first stage in a hybrid amplifier, see Fig.4.7.

Noise introduced by the first stage of a multistage amplifier is particularly important because it gets amplified by subsequent stages along with the signal. Besides producing the necessary gain, an amplifier must also produce a signal-to-noise ratio that is acceptable. Now, an f.e.t. is a low noise device thus it may be advantageous to use it in the first stage of an amplifier, especially when the amplitude of the input signal is small. Apart from this, an f.e.t. has a high input resistance and thus provides less loading of the signal source which is important when the signal source is of high internal resistance, i.e. the f.e.t. will provide a better match to a high impedance source such as a ceramic pick-up. Thus the circuit of Fig.4.7 may produce better overall results than the amplifier of Fig.4.5.

The f.e.t. TR1 uses a potential divider R1, R2 and source resistor R5 for biasing as described in Volume 2. Since R1, R2 would lower the input resistance, a stand-off resistor R3 is included which helps to decrease the effects of the potential divider on the signal source. The input signal $v_i$ is applied via C1 to the gate of TR1 and this common source amplifier produces an amplified signal voltage across R4. From R4 the signal is coupled via C3 to the base of TR2 operating as a common emitter amplifier. After amplification the output signal voltage $v_o$ is taken from across R9. The capacitors C1–C4 must all have a low reactance at the lowest frequency to be passed by the amplifier. As for the previous circuit only a.c. signals may be amplified.

### D.C. Coupled Amplifier

If the capacitors in series with the signal path are eliminated an amplifier is then able to amplify d.c. and very low frequencies, e.g. 1 Hz or below. A basic arrangement is shown in Fig.4.8 using two common emitter amplifiers in cascade. The coupling between TR1 and TR2 is now direct coupling or d.c. coupling. Thus the d.c. collector potential of TR1 becomes the base potential of TR2.

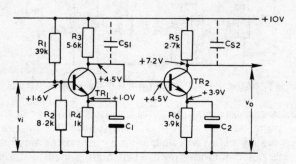

*Fig. 4.8 D.C. coupled amplifier.*

Assuming the same d.c. potentials on TR1 stage as in Fig.4.5, the base of TR2 will be at +4·5V as a result of the d.c. coupling. To provide the correct base-emitter voltage drop for TR2 of, say, 0·6V the emitter resistor of TR2 may be increased (now 3·9kΩ as opposed to 1kΩ in Fig.4.5) to give an emitter potential of +3·9V. This will give the required base-emitter voltage drop for TR2 and will maintain the same current in it as for Fig.4.5. To maintain the same collector-to-emitter voltage the value of TR2 collector load resistor will have to be reduced resulting in a higher collector potential for TR2 than for TR1. In a multistage amplifier, the collector potentials of successive stages will become larger and larger when this circuit arrangement is used. Also, the voltage gain of successive stages will reduce as the collector load resistor is reduced in value unless higher supply line voltages are used.

One of the problems of d.c. coupled amplifiers is d.c. drift. For example, if the collector current of TR1 changes, TR1 collector potential and hence TR2 base potential will change. This will result in a greater change in TR2 current and its collector potential, i.e. TR2 amplifies the d.c. drift. Since the steady operating conditions of TR2 have changed it may not operate satisfactorily. In Fig.4.8 d.c. drift is reduced by the emitter resistors R4 and R6 which introduce negative feedback since C1 and C2 are effectively open circuit to d.c. A variety of circuits have been designed to reduce d.c. drift, some employing differential amplifiers which have a low drift characteristic. These amplifiers that reduce effects of variations in temperature, supply voltage and transistor characteristics will be dealt with at a later stage in your course.

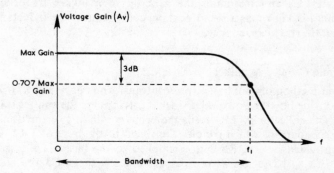

Fig. 4.9 Gain frequency response of D.C. coupled amplifier.

The d.c. coupled amplifier of Fig.4.8 will amplify both a.c. and d.c. signals. Its gain-frequency response is shown in Fig. 4.9. The gain is now maintained down to d.c. (0Hz), but in a practical amplifier there will be some departure from level response at low frequencies and d.c. At some high frequency the gain will start to fall off due to the stray capacitances of the circuit ($Cs_1$ and $Cs_2$) whose reactances will be low enough to shunt the collector loads appreciably. The bandwidth of the amplifier is from 0 Hz to $f_1$ where the gain has fallen to 0·707 of its steady level.

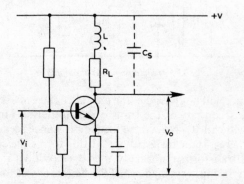

Fig. 4.10 Wide band amplifier 0Hz (D.C.) to 5.5 MHz.

To maintain the voltage gain up to a high frequency so that $f_1$ corresponds to, say, 5·5 MHz as in a t.v. video amplifier or 10 MHz for an oscilloscope or pulse amplifier, special circuit techniques are used to reduce the effect of stray capacitance. One idea is to use a peaking coil in series with the collector load resistor as in Fig.4.10. The load resistance $R_L$ is kept as small as is practical and at low and medium frequencies it is the effective load as the reactance of the peaking coil L is small. As the frequency is raised, the rising reactance of L increases the effective load impedance which compensates for the falling reactance of Cs and maintains the gain up to a higher frequency. This arrangement is known as a shunt peaking coil circuit as the inductance is in shunt with the stray capacitance.

### Transformer Coupled Amplifier

Another method of coupling is to use a transformer as in Fig.4.11. Bias for TR1 is provided by the potential divider R1, R2; R3 is the emitter stabilising resistor. The collector of TR1 feeds the primary winding of the transformer T1, the secondary of which couples the signal to the base of TR2. Bias for TR2 is provided by R4, R5 and is supplied to the base via the secondary winding. C3 by-passes R5 as regards a.c. since without C3 the secondary current would have to flow in R5 and produce a signal voltage drop across it which would cause a loss.

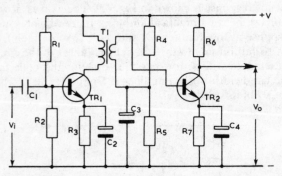

Fig. 4.11 Transformer coupled amplifier.

As a transformer will not pass d.c. the amplifier can only be used to amplify a.c. signals. The secondary of the transformer feeds the input resistance of TR2 which we will call $R_i$. Now the effective resistance seen at the primary is equal to $n^2R_i$ (where n is the turns ratio) and this forms the collector load for TR1 stage. The transformer normally uses a step-down ratio (primary-to-secondary) when the input resistance of TR2 stage is low as this gives increased voltage gain for TR1 stage. Usually, the step-down ratio is limited from about 3:1 to 4:1.

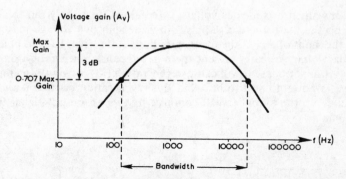

Fig. 4.12 Gain frequency response of transformer coupled amplifier.

The gain-frequency response for a transformer coupled stage is given in Fig.4.12. At low frequencies the gain drops due to the falling reactance of the transformer primary winding which reduces the load of TR1 stage. At high frequencies the fall in gain is caused by the stray circuit capacitances as for the previous circuits. Unless the circuit is well designed, the response is inferior to that obtained with capacitor coupling and there may be unwanted peaks at high frequencies due to resonant effects between the transformer inductance and secondary load capacitance. Transformer coupling is not common in small signal voltage amplifiers but is used in power amplifiers.

### Supply Line Decoupling

Consider the arrangement of cascaded amplifiers in Fig.4.13 where $i_1$, $i_2$ and $i_3$ represent the signal components flowing in the output of the amplifiers. These signal currents all flow in the internal resistance R of the d.c. supply.

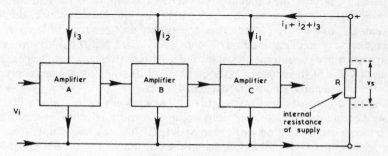

Fig. 4.13 Need for decoupling.

The internal resistance of the supply whether it be a battery or a mains unit is finite, thus there will be an unwanted voltage drop $V_s$ across it of magnitude $R(i_1 + i_2 + i_3)$.

This unwanted or spurious voltage is fed along the supply rail to all three stages together with the d.c. supply. Now the spurious voltage may find its way to the input of any of the three amplifiers. If it arrives back in phase with the signal voltage already present there it will cause regeneration or oscillation, but if in antiphase it will cause degeneration (loss of gain). In amplifiers, the most important effect to be avoided is regeneration because if an oscillation builds up the amplifier will continue to give an output signal with no input signal.

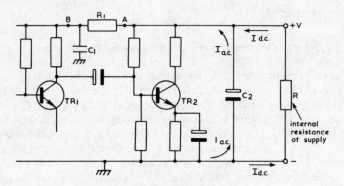

Fig. 4.14 Decoupling network in supply line.

Two measures that can be taken to reduce the amount of spurious voltage fed back along the supply rail are: (a) reduce the internal resistance of the supply to signal currents; and (b) introduce a decoupling network. Both are incorporated in Fig.4.14.

The internal resistance of the supply to the signal currents may be reduced by placing a large value capacitor (C2) across the supply. If the reactance of C2 at low frequencies is much smaller than the internal resistance, the a.c. currents $I_{ac}$ will be by-passed from the supply and produce little spurious voltage across it. This method is commonly used with battery-operated equipment since the internal resistance of a battery rises as it discharges. The value of C2 is of the order of 2000–3000$\mu$ F. With a mains power supply C2 is formed by the final smoothing capacitor.

There is a limit to the effectiveness of C2 since doubling its value will only halve its reactance and the spurious voltage. A more effective method is to introduce a decoupling network such as R1, C1 into the line supply feed, an arrangement used with all amplifiers. If the reactance of the shunt C1 at low frequencies is made small compared with the series R1, then most of the spurious voltage appearing at A will be dropped across R1 and very little will appear across C1, i.e. at B. Thus the amount of spurious voltage fed back along the line is reduced. The larger the value of R1 the better, but its value is often restricted especially in low voltage supplies due to the voltage drop across it produced by the steady current of TR1 and any earlier stages. In an audio amplifier R1 may be $100$–$200\Omega$ and C1 about $100$–$200\mu$F. The same arrangement is used in r.f. amplifiers but C1 can be made smaller, say, $0.001\mu$F to $0.1\mu$F depending upon the operating frequencies.

## Negative Feedback

Controlled feedback is used extensively in modern amplifier circuits by feeding back a fraction of the output signal to the input. If the fraction fed back is in antiphase with the input signal it is called negative feedback (n.f.b.).

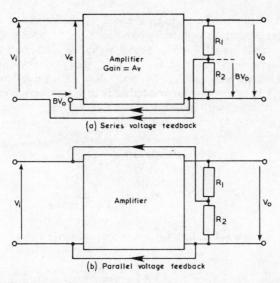

Fig. 4.15 Voltage feedback (feedback voltage proportional to output voltage).

Whenever n.f.b. is applied in an amplifier, the gain is reduced. There would, however, be little point in reducing the gain of an amplifier if there were not benefits to be gained from the use of n.f.b. The main benefits are:

(1) The stability of the amplifier is improved, i.e. its gain remains more stable in spite of variations in the values of both active and passive circuit components with time and temperature or after component replacement and d.c. supply variations. This is important particularly with amplifiers where the gain must remain stable over long periods, e.g. in test instruments.

(2) Distortion or noise introduced by the amplifier is reduced.

(3) The frequency response of the amplifier may be improved.
(4) Input and output resistances of the amplifier may be altered to provide better matching.

There are different methods used for deriving the feedback signal and applying it to the input which will now be considered.

### Series and Parallel Voltage Feedback

With series and parallel voltage feedback, the fraction of the signal fed back in antiphase is proportional to the output voltage of the amplifier. The idea of series voltage feedback is shown in Fig.4.15(a).

### Effect on Gain

A fraction B of the output voltage, decided by the values of R1 and R2 connected across the output voltage $V_o$ is fed back to the input and applied in series with the input signal $V_i$. The actual signal input to the amplifier is the difference between $V_i$ and the feedback signal (since they are in series opposition) which we will call $V_e$. Suppose that the gain of the amplifier without feedback is $A_v$ and the fraction fed back in antiphase is $BV_o$.

$$\text{Now, with feedback } V_e = V_i - BV_o$$
$$\text{and } V_o = A_v V_e$$
$$\therefore V_o = A_v(V_i - BV_o)$$
$$\text{or } A_v V_i = V_o + A_v BV_o$$
$$\text{or } A_v V_i = V_o(1 + A_v B)$$

$$\text{or } \frac{V_o}{V_i} = \frac{A_v}{1 + A_v B}$$

but $\dfrac{V_o}{V_i}$ is the gain with feedback $(A_v')$

$$\therefore A_v' = \frac{A_v}{1 + A_v B}$$

Although the proof of the expression for gain has been shown, it is the general result that is probably more important. Consider the following example.

The voltage gain of an amplifier without feedback is 500. What will be its gain if n.f.b. is used when the fraction fed back is $\frac{1}{100}$?

$$\text{Now } A_v' = \frac{A_v}{1 + A_v B}$$

$$= \frac{500}{1 + (500 \times \frac{1}{100})}$$

$$= \frac{500}{6}$$

$$\simeq 83.$$

## Effect on Distortion

If an amplifier without feedback introduces a percentage distortion D, then when n.f.b. is applied the distortion is reduced by the factor $1 + A_vB$. The expression for the new distortion D′ with feedback can be derived in a similar way to that given for gain but only the result will be given:

$$D' = \frac{D}{1 + A_vB}$$

Consider the application of the result to the following example.

An amplifier without feedback has a gain of 200 and introduces $10\%$ distortion. What will be the distortion when n.f.b. is used and the fraction fed back is $\frac{1}{100}$?

$$\text{Now } D' = \frac{D}{1 + A_vB}$$

$$= \frac{0 \cdot 1}{1 + (200 \times \frac{1}{100})}$$

$$= \frac{0 \cdot 1}{3}$$

$$= 0 \cdot 033$$

$$= 3 \cdot 3\%$$

## Effect on Noise

Noise introduced by the amplifier is treated in the same way as distortion and is reduced by the factor $1 + A_vB$.

## Effect on Frequency Response

Suppose that without feedback the voltage gain of an R–C coupled amplifier at a frequency $f_1$ is 100 and that at a higher frequency $f_2$ the gain has fallen to 80 (due to the effect of stray circuit capacitance). If n.f.b. is used and the fraction fed back is, say, $\frac{1}{10}$ the new gains will be:

At $f_1$

$$A'_v = \frac{100}{1 + (100 \times \frac{1}{10})}$$

$$\simeq 9$$

and at $f_2$

$$A'_v = \frac{80}{1 + 80 \times \frac{1}{10}}$$

$$\simeq 8 \cdot 9$$

Thus without feedback the gain reduction from $f_1$ to $f_2$ is 20% of that at $f_1$, but with n.f.b. it is only 1.1.% of that at $f_1$, i.e. the frequency response has been improved (flatter response). This of course is at the expense of a reduction in gain.

### Effect on Stability

In dealing with the effect on frequency response, it was assumed that the gain of the amplifier reduced due to the effect of stray circuit capacitance. The same results are obtained whatever the cause of change in amplifier gain, e.g. if a transistor is replaced by another of the same type but with a different $h_{fe}$, or the supply line voltage varies. Consider the following example:

The voltage gain of an amplifier without feedback is 100 and on replacing a transistor it increases to 110 (10% increase). What will be the percentage increase if n.f.b. is used and the fraction fed back is $\frac{1}{5}$?

$$\text{Gain before replacing transistor} = \frac{100}{1 + (100 \times \frac{1}{5})}$$

$$= 4 \cdot 76$$

$$\text{Gain after replacing transistor} = \frac{110}{1 + (110 \times \frac{1}{5})}$$

$$= 4 \cdot 78$$

$$\% \text{ increase} = \frac{0 \cdot 02}{4 \cdot 76} \times 100$$

$$= 0 \cdot 42 \%$$

Thus the stability of the amplifier has been greatly inproved.

### Effect on Input and Output Resistance

With series voltage feedback, the input resistance $R_i$ can be shown to be increased to $R_i (1 + A_v B)$ and the output resistance $R_o$ decreased to:

$$\frac{R_o}{1+A_vB}$$

An alternative method of applying the feedback signal to the input is shown in Fig.4.15(b). A fraction of the output signal voltage appearing across R2 is fed back to the input and applied in parallel with the input signal. This method of applying the feedback signal which is proportional to the output voltage, reduces the gain and distortion and improves the frequency response and stability of the amplifier as for series voltage feedback. However, it decreases both the input and output resistance of the amplifier.

### Series and Parallel Current Feedback

With, so-called, current feedback a voltage is fed back to the input in antiphase with the input signal, but in this method the feedback voltage is proportional to the output current.

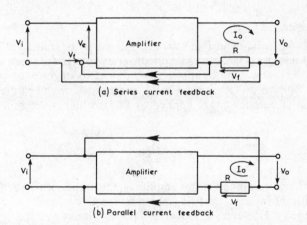

(a) Series current feedback

(b) Parallel current feedback

Fig. 4.16 Current feedback (feedback voltage proportional to output current).

The feedback signal may be applied in series with the input as in Fig.4.16(a) or in parallel with it as in diagram (b). To develop the feedback voltage $V_f$, a resistor R is used through which the output current $I_o$ of the amplifier is passed.

As far as gain, noise and distortion are concerned, current feedback has the same effects as for voltage feedback; also, the frequency response is improved if the amplifier load is resistive. For series current feedback both input and output resistance are increased whereas for parallel current feedback the input resistance is reduced and the output resistance increased. The details of the four methods of deriving and applying n.f.b. are sumarised in the table below.

|  | Effect of n.f.b. on | | | | | |
| Method | Voltage Gain | Stability | Distor-tion | Frequency response | Input resistance | Output resistance |
|---|---|---|---|---|---|---|
| Series Voltage | Reduced | Improved | Reduced | Improved | Increased | Reduced |
| Parallel Voltage | Reduced | Improved | Reduced | Improved | Reduced | Reduced |
| Series Current | Reduced | Improved | Reduced | Improved | Increased | Increased |
| Parallel Current | Reduced | Improved | Reduced | Improved | Reduced | Increased |

### N.F.B. Circuits

Ideas of how the basic feedback arrangements may be translated into n.f.b. circuits will now be given. An example of series voltage feedback is given in Fig.4.17(a). This shows n.f.b. applied overall to two stages of amplification. The output voltage of TR2 is fed via R1 to across R2 in the emitter of TR1. The fraction of output voltage applied to TR1 emitter is determined by the ratio:

$$\frac{R_2}{R_1 + R_2}$$

Now, with two common emitter amplifiers, the output of the second will be in phase with the input to the first (shown by the arrow heads pointing in the same direction). However, as regards TR1 the input voltage $V_i$ and the feedback voltage $V_f$ are in antiphase, since one is applied to the base and the other to the emitter. The effective input signal $V_e$ applied between the base and emitter of TR1 is equal to $V_i - V_f$ and is therefore less than the signal $V_i$ applied to the input terminals. Thus the effect of n.f.b. is to reduce the input signal and hence the output signal which means that the gain is reduced. There is also some local series current feedback on TR1 stage (see later).

An example of parallel voltage feedback is shown in Fig.4.17(b) where the feedback is from collector to base. The output voltage, which is in antiphase with the input over one stage of common-emitter amplification, is fed back to the input via C1 and R1. The capacitor C1 blocks the d.c. and will be assumed to have zero reactance at all frequencies. Since the d.c. is blocked only a.c. feedback is considered, as also in diagram (a). The amount of feedback is determined by the value of R1 and the resistance of the input circuit.

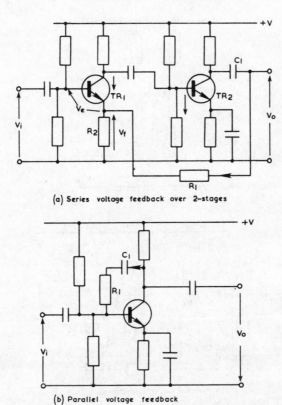

(a) Series voltage feedback over 2-stages

(b) Parallel voltage feedback

*Fig. 4.17 Circuits with voltage negative feedback applied.*

The most common way of achieving series current feedback is shown in Fig.4.18(a). Here an undecoupled resistor R1 is placed in the emitter circuit of the transistor. This resistor will carry the output current $I_o$ and thus there will be a signal voltage across it that is proportional to the output signal current. This voltage $V_f$ is the feedback voltage and it will act in opposition to the input signal $V_i$. Thus the effective signal $V_e$ applied between base and emitter is reduced since it is equal to $V_i - V_f$. Thus the voltage gain of the amplifier is reduced. The amount of feedback is settled by the ratio of $R_1/R_3$. The resistor R2 does not produce any n.f.b. to a.c since it is decoupled by C1. If C1 were omitted the n.f.b. would be increased and the gain reduced to a greater degree. The presence of R1 will affect the steady d.c. at the emitter.

An arrangement for parallel current feedback is given in Fig.4.18(b). This shows overall feedback applied over two stages of amplification. The voltage to be fed back is developed across the emitter resistor R1 of TR2 which carries the output current $I_0$. The signal voltage across R1 will be in antiphase with the input signal voltage $V_i$ since it is of the same phase as

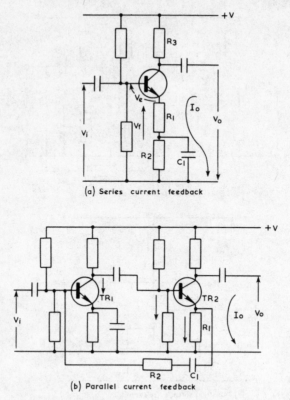

(a) Series current feedback

(b) Parallel current feedback

*Fig. 4.18 Circuits with current negative feedback applied.*

TR2 base and TR1 collector. After passing through C1 (assumed to be of zero reactance at all frequencies) which blocks the d.c. and R2 the feedback signal is applied in parallel with the input circuit of TR1. The degree of feedback is determined by the values of R2 and the input resistance of TR1.

One of the circuit configurations of the ordinary transistor and the f.e.t. which relies on n.f.b. for its special characteristics is the emitter follower or source follower (also known as common collector and common drain), see Fig.4.19. These circuits employ 100% series voltage feedback, since the full output voltage $V_o$ across the emitter or source resistor acts in phase opposition to the input voltage $V_i$ and reduces the effective input voltage $V_e$ to a very small value. This results in (a) a voltage gain less than unity; (b) a high input resistance; and (c) a low output resistance. These features allow the circuits to be used to good advantage in matching a high resistance source to a low resistance load, i.e. to act as a buffer stage.

In practice the high input resistance of the emitter follower circuit may be limited due to the presence of the biasing resistors R2 and R3. When a very

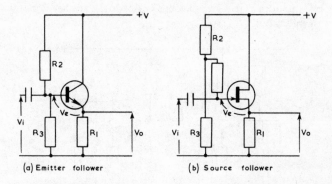

*Fig. 4.19 Emitter and source follower circuits (series voltage feedback).*

high input resistance is required, the circuit may be altered to that shown in Fig.4.20 with components R4 and C1 added. If the reactance of C1 is considered to be zero at all frequencies, the signal voltage at point B will be the same as the emitter-signal voltage. The signal voltage at point A is the input voltage which is applied between base and the negative supply line. Since the base and emitter signal voltages are approximately the same, the signal voltage across R4 will be very small, i.e. the signal current in R4 will be very small. Thus the effect is that the input signal hardly sees the potential divider R1, R2, or its resistance to the input signal is increased due to the presence of R4 and C1.

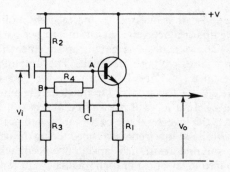

*Fig. 4.20 A Bootstrapped emitter follower.*

## Frequency Selective n.f.b.

In some applications it is required to make the gain of an amplifier increase or decrease over a particular section of its frequency band. This can be achieved by including in the feedback network reactive elements, i.e. capacitors or inductors.

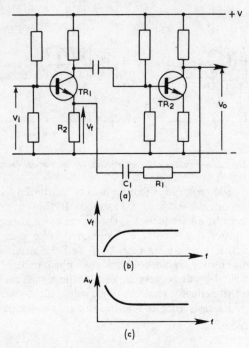

Fig. 4.21 Frequency selective N.F.B.

For example, suppose it is desired to increase the voltage gain at low frequencies to make up for a deficiency at l.f. in another part of an electronic system. Fig.4.21 (a) shows a possible circuit arrangement for obtaining the desired effect using series voltage feedback. The amount of feedback is settled by the values of R1, C1 and R2. If a value for C1 is chosen so that it produces a rising reactance towards low frequencies the amount of feedback voltage $V_f$ developed across R2 will rise as in diagram (b). The voltage gain Av of the amplifier is the inverse of the curve for $V_f$ and is shown in diagram (c). Thus the gain will rise at low frequencies producing the desired l.f. boost.

By using other suitable resistance-capacitance feedback networks, low frequency cut, high frequency boost or high frequency cut can be achieved.

## D.C. Feedback

When overall n.f.b. is used with a d.c. coupled amplifier such as that shown in Fig.4.22, the feedback can be made to assist in stabilising the d.c. operating conditions of the amplifier as well as operating on the signal input.

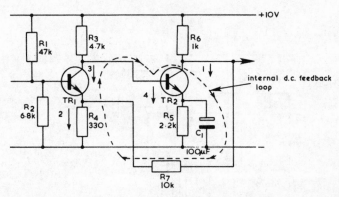

*Fig. 4.22 Amplifier with A.C. and D.C. feedback.*

As there is no d.c. blocking capacitor between the collector of TR2 and the emitter of TR1, d.c. variations at TR2 collector are fed back to TR1 emitter. There is a d.c. feedback loop formed (shown dotted) which will assist in correcting d.c. changes within the loop. Suppose, for example, that TR2 collector current increases as a result of a rise in the ambient temperature causing a fall in TR2 collector voltage (indicated by the arrow marked 1). A fraction of this fall in voltage is applied to TR1 emitter (indicated by the arrow 2). Now, a fall in emitter voltage will result in TR1 collector current increasing and its collector voltage falling (indicated by arrow 3). The fall of TR1 collector voltage is applied to TR2 base (indicated by arrow 4). This will cause TR2 collector current to fall and for its collector voltage to rise which will oppose the original fall. Thus the d.c. feedback will help to stabilise the d.c. operating conditions. Other examples of d.c. feedback will be considered in Chapter 5 on power amplifiers.

### I.C. Amplifier

Linear amplifiers, using similar principles to the discrete transistor circuits already described, are available in integrated circuit form and are in common use. Such i.c.s. can be designed to cover a wide range of frequencies and applications and are normally based on bipolar transistor technology. Because of the difficulty of fabricating large value capacitors in i.c. form, capacitor coupling is dispensed with to give way to direct coupling between stages within the i.c. Thus any large value coupling or decoupling capacitors that may be required have to be fitted externally to the i.c. as are any large value resistors that may be needed.

A linear i.c. amplifier will often include a mixture of common emitter, emitter follower and differential amplifier stages. Some n.f.b. is incorporated in most linear i.c.s. to provide stability of gain and d.c. stabilisation and provision is often made for additional feedback to be applied externally.

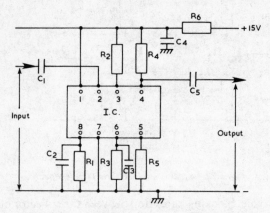

*Fig. 4.23 Integrated circuit amplifier.*

A typical form for an integrated circuit amplifier is shown in Fig.4.23, where a combination of discrete circuit components and an integrated circuit is used. The capacitor C1 couples the input signal to the i.c. on pin 2 and the capacitor is used to block the d.c. component to and from the i.c. Capacitor C5 serves a similar purpose for coupling the output signal from pin 4. The external d.c. supply is fed to pins 1 and 7 and this provides the basic d.c. supply to the various amplifier stages within the i.c. Resistors R1 and R3 are emitter resistors which are suitably decoupled by C2 and C3. Resistors R5 and R2 are bias resistors which do not require decoupling. Resistor R4 may serve as a collector load resistor for one of the amplifier stages. R6 and C4 provide a line decoupling network to prevent instability in the i.c. The values of the coupling and decoupling capacitors will depend upon the frequencies of the signals to be amplified; e.g. for an audio amplifier these capacitors will be in the range of, say, $1\mu$F to $50\mu$F.

### Tuned Amplifiers

A tuned amplifier is used where it is desirable to amplify a relatively narrow band of frequencies to the exclusion of all other frequencies. Thus a tuned amplifier has to provide gain and selectivity. The gain may be provided by a discrete transistor stage or i.c. and the selectivity by a tuned circuit or ceramic filter.

A basic tuned amplifier stage is given in Fig.4.24(a) using a bipolar transistor. The collector load of the amplifier is a parallel resonant circuit L1, C1 which is tuned to the centre of the band of frequencies to be amplified. Forward biasing of the transistor to class-A operation is provided by R1, R2 with R3 serving as the emitter stabilising resistor. Capacitor C3 decouples

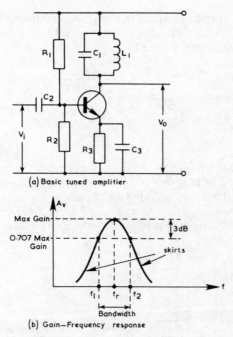

(a) Basic tuned amplifier

(b) Gain—Frequency response

Fig. 4.24 Tuned voltage amplifier.

R3 to prevent n.f.b.

At the resonant frequency $(f_r)$ the impedance of a parallel tuned circuit is high and purely resistive. The impedance at resonance is sometimes called the dynamic impedance $R_D$ and is equal to:

$$\frac{L}{Cr} \text{ where r is the tuned circuit losses.}$$

The voltage gain of a common emitter amplifier is given by the expression:

$$Av = \frac{h_{fe} \times R_L}{R_{IN}}$$

With a tuned amplifier, the load at resonance is the dynamic impedance $R_D$, thus the expression for voltage gain is

$$Av = \frac{h_{fe} \times R_D}{R_{IN}}$$

At the resonant frequency $R_D$ is high and the voltage gain can be quite large. Away from resonance the impedance of the tuned circuit falls and thus the gain of the amplifier falls. The response of the amplifier therefore takes the form shown in Fig.4.24(b). The bandwidth of a tuned amplifier is taken to be between the frequency limits where the gain has fallen to 0·707 of its maximum value, i.e. 3dB down from maximum gain. When the circuit losses are small, the resonant frequency is given by

$$f_r = \frac{1}{2\pi\sqrt{L.C}}$$

(f in Hz, L in henries and C in farads)

### Tuning

Some tuned amplifiers use fixed tuning, i.e. the frequency of operation is set by the manufacturer or during servicing, but in normal use the frequency is not varied. This type is found in i.f. amplifiers in receivers and in other signal applications. On the other hand, r.f. amplifiers in receivers use variable tuning so that the receiver may be set to receive a transmission on a different operating frequency.

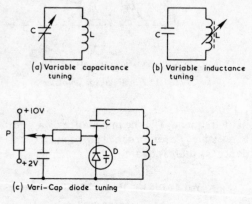

(a) Variable capacitance tuning

(b) Variable inductance tuning

(c) Vari-Cap diode tuning

*Fig. 4.25 Tuning methods.*

With variable tuning, the resonant frequency may be altered by using a variable tuning capacitor or by using a variable inductor (one with an adjustable ferrite core). Alternatively, a vari-cap diode may be included in the tuned circuit so that it forms part of the circuit capacitance. Tuning is then altered by applying a variable d.c. to vary the reverse bias of the diode and hence its capacitance. These tuning methods are illustrated in Fig.4.25.

### Obtaining the Required Bandwidth

If the band of frequencies to. be amplified is small, we require a tuned circuit of high selectivity or high Q factor. A high Q factor may be ensured by making the ratio:

$$\frac{\text{inductive reactance of tuning coil}}{\text{tuned circuit losses}} \qquad \frac{2\pi f L}{r}$$

large and providing a narrow bandwidth. On the other hand if a compara-tively wide bandwidth is required a tuned circuit of low Q could be used. However, a low Q tuned circuit has skirts that are not very steep and the selectivity may be inadequate. Thus to obtain a wide bandpass with good selectivity alternative methods are used.

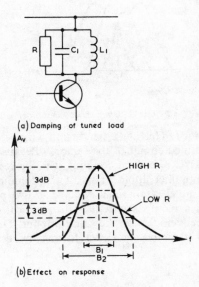

(a) Damping of tuned load

(b) Effect on response

Fig. 4.26 Effect of damping.

One method is to use a tuned circuit of high Q and to damp it with a resistor R as shown in Fig.4.26(a). The effect of damping is shown in diagram (b). As the resistance of R is decreased, the damping is increased providing an increase in the 3dB bandwidth. The presence of R in the collector circuit of the transistor reduces the value of the effective load, so the voltage gain is less. If the gain is insufficient, further stages of amplification will be necessary.

Better selectivity can be obtained by using double tuned circuits as shown in Fig.4.27(a). Here the two tuned circuits L1, C1 and L2, C2 of the same Q and made resonant to the same frequency are coupled together via mutual inductance M. The actual response obtained will depend upon the degree of coupling between the two tuned circuits, see diagram (b). With weak coupling the response at A is obtained and is similar to that of a single tuned circuit. If the coupling is increased to what is termed critical coupling the flat-topped

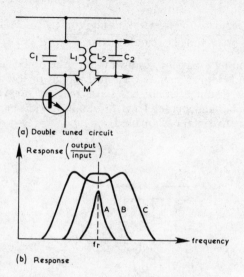

(a) Double tuned circuit

(b) Response

Fig. 4.27 Use of bandpass coupled tuned circuits.

response at B is obtained. This has a bandwidth which is increased by 1·414 over that of a single tuned circuit of the same Q. Because of the flat-topped response and steep skirts of critical coupling it is the response that is often aimed for in tuned amplifier applications. If the coupling is increased still further, the response at C is obtained which has two peaks. This type of circuit is often called a bandpass circuit; when critical coupling is used it amplifies a band of frequenceies by the same amount. The bandwidth of the double tuned circuit may be increased further. by connecting damping resistors across it.

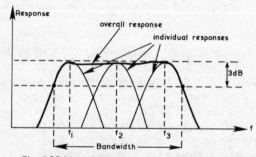

Fig. 4.28 Idea of stagger tuning to increase bandwidth.

When a very wide bandwidth is required, a technique known as stagger tuning may be employed. This is used when a number of tuned amplifier stages are cascaded and the tuned circuit(s) of each stage are tuned to different frequencies within the band of frequencies to be amplified. The idea of this technique, illustrated in Fig.4.28, results in an overall wide bandwidth with good selectivity.

Some examples of practical tuned amplifier circuits will now be considered under two sections: variable tuned amplifiers and fixed tuned amplifiers.

### Variable Tuned Amplifiers

A variable tuned amplifier is the type used in the r.f. stages of a receiver One example is given in Fig.4.29 which would be suitable for a receiver operating around 100 MHz such as a broadcast f.m. receiver. The circuit is based on a common emitter amplifier TR1 with tuned circuits connected at input and output.

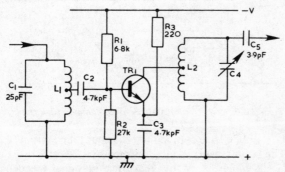

*Fig. 4.29 R.F. amplifier (100MHz) using common emitter circuit.*

The signal to be amplified is developed across the input tuned circuit L1, C1. This is a fixed tuned circuit having a bandwidth which embraces the full frequency range of the receiver. From the input tuned circuit the signal is coupled via C2 to the base of TR1. This capacitor acts as a d.c. block. Resistors R1 and R2 provide forward bias to class-A and R3 is the emitter stabilising resistor. At first it may appear that the resistors R1 and R2 have been inadvertently interchanged, but this is in order as the circuit is using an upside down supply. Capacitor C3 decouples R3 to a.c. to prevent n.f.b. (note that as far as a.c. is concerned the negative and positive supply rails are at the same potential).

The load for the transistor is the parallel tuned circuit L2, C4 with variable tuning provided by C4. The 'upside down' supply arrangement allows one side of the variable tuning capacitor C4 to be connected to the neutral chassis line. The output from the stage is fed via a small value capacitor C5 to the next stage (normally the mixer). To prevent the input and output resistances of TR1 damping the tuned circuits unduly, the input to TR1 is obtained from a tap on L1 and the output from TR1 is fed to a tap on L2. The input and output resistances of TR1 are present between the taps on the inductors and the chassis line. Therefore, due to auto-transformer action, they will appear as larger resistances across the full tuned circuits.

More usually the r.f. stage of a receiver operating at high frequencies uses the common base configuration. A common base circuit gives a better frequency response than a common emitter circuit for a given transistor but has the disadvantage of a lower input resistance, Fig.4.30.

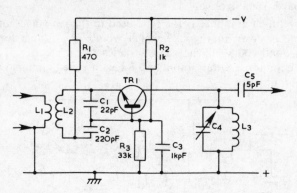

*Fig. 4.30 R.F. amplifier (100MHz) using common base circuit.*

Signals to be amplified are coupled to the input tuned circuit via a coupling winding L1. The input tuned circuit is formed by L2 and C1 (the circuit being completed via C2 which has a low reactance at signal frequencies). As the input resistance of TR1 is very low (tens of ohms) there is little point in making the tuning of the input circuit variable, so fixed tuning is used. The selectivity of the input circuit is therefore limited because of the heavy damping.

Base bias is provided by the potential divider R2, R3 and C3 decouples the base to the chassis line, i.e. the base is at neutral potential as regards a.c. (common base). Again, an upside down supply is used which allows one side of C4 to be connected to the chassis line. Resistor R1 is the emitter d.c. stabilising resistor which is decoupled by the series combination of C2 and C3.

The collector load for TR1 is the parallel tuned circuit L3, C4 with variable tuning provided by C4. The output from the circuit is fed via the small value capacitor C5.

An f.e.t. may be used as an r.f. amplifier where its special features of low noise, higher input resistance and good signal handling are advantageous. A circuit illustrating modern practice is given in Fig.4.31.

TR1 is a depletion mode mosfet, bias to class-A being provided by the source resistor R2 which is decoupled by C4 to prevent n.f.b. Additionally, the bias is augmented by an a.g.c. voltage applied via R1 to the gate electrode to deal with variations in signal strength. Capacitor C3 decouples the a.g.c. line to prevent instability.

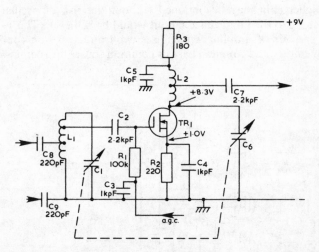

*Fig. 4.31 R.F. amplifier (100MHz) using F.E.T. in common source circuit.*

Signals to be amplified are applied via C8 and C9 to the input tuned circuit comprising L1 and C1. Variable tuning of the input circuit is provided by C1 which improves the overall selectivity of the stage. Signals from the tuned circuit are coupled to the gate of TR1 by C2 which serves as a d.c. block for the a.g.c. voltage. The gate tap on L1 allows correct matching between the input impedance of TR1 and the input tuned circuit, thereby maintaining adequate Q and selectivity (note that the f.e.t. input impedance is only several kilohms at 100 MHz due to its input capacitance). The drain load for TR1 is the parallel tuned circuit L2, C6 with C6 providing variable tuning. Note that for a.c. C6 is effectively in parallel with L2 since the upper end of L2 is connected to the chassis line via C5. This capacitor together with R3 forms a supply line decoupling filter. Note also the smaller values of the decoupling capacitors C3, C4 and C5 at these higher frequencies. The output signal is coupled to the following stage via C7 from a tap on L2 to prevent undue damping of the output tuned circuit. Capacitors C1 and C6 are ganged for ease of tuning.

With a common-source or common-emitter circuit there is some capacitance between the output and input electrodes of the transistor. Although this capacitance is small (a few picofarads) it will allow feedback to occur at high frequencies which may cause the circuit to oscillate. When a common gate or common base circuit is used the capacitance between the output and input electrodes of the transistor is considerably reduced which lessens the risk of oscillation. However, with common gate or common base circuits the lower input resistance causes damping of the input tuned circuit. If the low feedback capacitance of the common gate or common base circuit could be combined with the higher input resistance of the common-source or common-emitter circuit a higher gain may be obtained without the risk of oscillation and better matching to the input tuned circuit would be achieved. This is the idea behind the cascode r.f. amplifier which uses two transistors in a combination of common emitter and common base or common source and common gate.

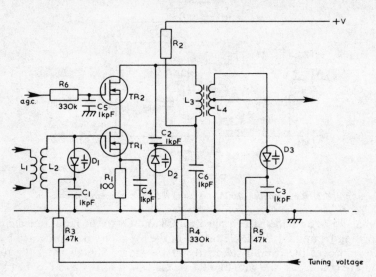

*Fig. 4.32 Cascode R.F. amplifier (100MHz) using F.E.Ts. and electronic tuning.*

A cascode r.f. amplifier using f.e.ts. is given in Fig.4.32. TR1 and TR2 form the cascode stage with TR1 connected in common source and TR2 in common gate. As far as d.c. is concerned the two transistors are connected in series across the d.c. supply with the drain current of TR1 being the source current for TR2.

Electronic tuning is used in this circuit with three variable tuned circuits formed by L2, D1; L3, D2; and L4, D3. The vari-cap diodes D1-D3 are all fed from a common tuning line voltage via decoupling networks R3, C1 and R5, C3.

The signal to be amplified is fed to the input tuned circuit L2, D1 via a coupling winding L1. From the tuned circuit the signal is applied to the gate of TR1. Bias to class-A is provided by the source resistor R1 which is suitably decoupled by C4 to prevent n.f.b. Since the damping produced by the common source circuit is small the input circuit will have good selectivity. TR1 drives the upper transistor TR2 with the low input impedance of TR2 acting as the load for TR1, thus the gain of TR1 stage is small. Most of the voltage gain is produced by the common gate stage TR2. Capacitor C5 places TR2 in common gate by connecting the gate to chassis for signals. TR2 gate electrode could be fed from a potential divider so that the transistor is biased correctly but in this circuit an a.g.c. voltage of positive polarity is fed to the gate to deal with varying signal strengths.

TR2 load is formed by the parallel tuned circuit L3, D2 which together with L4, D3 form a bandpass coupled circuit thus improving selectivity. The d.c. supply is fed to the drain of TR2 via the decoupling filter R2, C6 and L3. Capacitors C1, C2 and C3 act as d.c. blocking capacitors for the tuning voltage. These capacitors have a low reactance to signal frequencies and complete the a.c. circuit in their respective tuning circuits.

### Fixed Tuned Amplifiers

Fixed tuned amplifiers are used in receivers for i.f. amplification. These amplifiers provide most of the receiver gain which necessitates the provision of two or more stages of i.f. amplification. One example is given in Fig.4.33 which illustrates the method of coupling between cascaded i.f. stages.

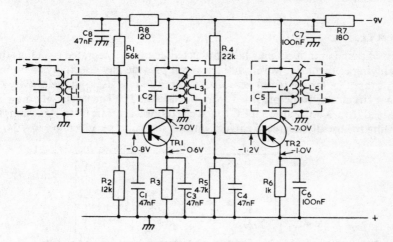

*Fig. 4.33 Cascaded I.F. stages (470KHz).*

The i.f. signal to be amplified is applied from a coupling winding L1 to the base of TR1 operating in common emitter and using a p-n-p transistor. Correct bias is provided by R1, R2 with R3 acting as the emitter d.c. stabilising resistor and suitably decoupled by C3. Capacitor C1 by-passes R2 at i.f. to prevent loss of signal across it. The collector load for TR1 is formed by the parallel tuned circuit L2, C2 which is tuned to the i.f. of the receiver (470 kHz in an a.m. radio receiver). TR1 collector is fed to a tap on L2 to prevent the output resistance of TR1 excessively damping the tuned circuit to maintain the selectivity.

Coupling to the base of TR2 is via a winding L3 which provides a step-down transformer action with L2. This is necessary to match to the following stage and thus prevent the comparatively low input resistance of TR2 damping

L2, C2 unduly. The components C2, L2 and L3 are always enclosed in a screening can, the assembly being referred to as an i.f. transformer. The p-n-p transistor TR2 amplifies the signal applied to its base and this common-emitter stage feeds the collector load consisting of the parallel tuned circuit L4, C5. Base biasing for TR2 is provided by R4, R5 with C4 decoupling R5 to prevent loss of i.f. signal. Resistor R6 provides d.c. stabilisation and is suitably decoupled by C6 to prevent n.f.b.

The output to the following stage is coupled by the winding L5 which, together with L4, forms a further i.f. transformer. The d.c. supply line is filtered by R7, C7 and R8, C8. Typical voltages are given with the base-emitter voltage drops (0·2V) being appropriate to germanium type transistors.

Sometimes double-tuned circuits are used but these require tapping at both primary and secondary to prevent loss of selectivity. Inductors L2 and L4 have adjustable cores so that the tuned circuits may be aligned to the intermediate frequency.

## Use of i.c.

Fixed tuned amplifiers may be constructed using an integrated circuit as the amplifying element. However, an i.c. cannot provide any selectivity as tuning coils cannot be fabricated in i.c. form. When an i.c. is used, the selectivity or tuning circuit has to be provided from outside the i.c. The selectivity may be obtained from a conventional LC tuned circuit or a ceramic filter. An example showing the use of integrated circuits with ceramic filters is given in Fig.4.34.

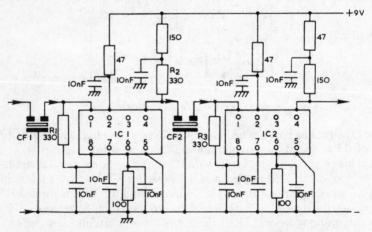

Fig. 4.34 I.F. amplifier (10.7MHz) using integrated circuits and ceramic filters.

A ceramic filter makes use of the piezo-electric property of certain processed ceramic materials. This property is the same as found in quartz crystals which are discussed in Chapter 6. The special construction of a ceramic filter allows bandpass properties to be achieved providing good selectivity, but its centre frequency is not adjustable. When more than one filter is used it is important that they are all tuned to the same centre frequency. Thus the filter units are graded by the manufacturer into frequency groups and colour coded. The filters used in any i.f. amplifier must be from the same frequency group.

The ceramic filters in Fig.4.34 are CF1 and CF2 (10·7 MHz centre frequency) and they provide the selectivity of the i.f. amplifier. Integrated circuits IC1 and IC2 produce the required gain with discrete components added externally to provide biasing, decoupling and loads, etc. as previously explained. The filters have to be matched at input and output; resistors such as R2 and R3 (330Ω) are used for this purpose.

### Tuned Amplifier Responses

Typical responses for some tuned amplifier stages are illustrated in Fig.4.35. In diagram (a) we have the response for the r.f. stage of an f.m. radio receiver such as that shown in Fig.4.32. Diagram (b) gives the response for the i.f. stages of a radio receiver such as that given in Fig.4.33. Finally, the response for the i.f. stages of an f.m. radio receiver such as those shown in Fig.4.34 is given in diagram (c).

It should be mentioned that when tuned amplifiers are cascaded with their tuned circuits resonant at a common centre frequency, there is a bandwidth shrinkage. Thus the 3dB bandwidth of the individual stages will be greater than the overall 3dB bandwidth.

### Methods of Achieving A.G.C.

Automatic gain control (a.g.c.) is used in receivers to offset the effects of variations in the received signal strength. With a radio receiver the object is to provide a constant sound volume level when stations of differing signal strengths are tuned-in. With a television receiver the object is to obtain a picture of constant contrast level when tuning to stations operating on different channels (also the sound level is kept constant).

It is not intended here to deal with methods of deriving a control voltage, but to show how automatic gain control of a receiver may be achieved. With a bipolar transistor there are two ways of altering its gain (discussed in Volume 2): reverse gain control and forward gain control.

We will first consider reverse gain control and an example is given in Fig.4.36. Gain control is carried out in the r.f. and/or the i.f. stages of a receiver but as an a.m. radio receiver does not usually employ an r.f. stage, control can only be applied in the i.f. stages (usually the first i.f. stage or the first two i.f. stages). To reduce the gain of TR1, the forward bias is reduced (reverse gain control). The value of the bias used on TR1 is settled by the the values of R1, R2 and the magnitude of the a.g.c. voltage applied to R2.

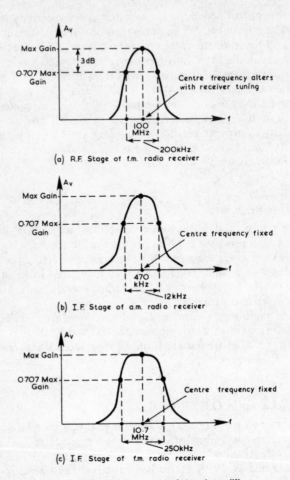

(a) R.F. Stage of f.m. radio receiver

(b) I.F. Stage of a.m. radio receiver

(c) I.F. Stage of f.m. radio receiver

Fig. 4.35 Typical responses of tuned amplifiers.

Under weak signal conditions it will be assumed that the d.c. voltages are as shown in the diagram thus producing a forward bias of 0·2V (germanium transistor). If the signal strength increases, the magnitude of the a.g.c. voltage, which is made proportional to signal strength, increases which causes the voltage at point A to move in a positive direction. As a result the forward bias on the transistor is reduced and the voltage gain of the stage is lowered. Thus an increase in signal strength causes the gain to be reduced which tends to maintain a constant output signal from the receiver. If the signal strength is now decreased, there will be less a.g.c. voltage and point A will move in a negative direction causing an increase in the bias on TR1. As a result the voltage gain of the i.f. stage will be increased thereby compensating for the reduction in signal strength. R2, Cl form a decoupling filter for the a.g.c. line.

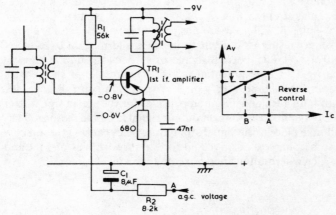

Fig. 4.36 Reverse gain control in I.F. stage of A.M. radio receiver.

The gain control used in television and f.m. radio receivers is usually of the forward type as there are disadvantages in using reverse control in these receivers. Since television and f.m. radio receivers employ an r.f. stage, a.g.c. may be applied to it in addition to control on the i.f. stages. An example of forward control on the i.f. stages of a television receiver is shown in Fig.4.37. Here the control operates on both TR1 and TR2 stages. TR2 base obtains its potential from the a.g.c. line voltage which we will assume to be approximately +6·7V under weak signal conditions. Now, the emitter voltage of TR2 will be its base voltage less its base-emitter voltage drop of, say, 0·6V (silicon transistor). Thus the emitter potential will be +6·1V as

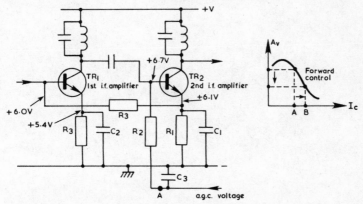

Fig. 4.37 Forward gain control in I.F. stages of television receiver.

shown. TR1 receives its base potential from TR2 emitter, so TR1 base voltage will be approximately that of TR2 emitter (there will be some voltage drop across R3 due to TR1 base current). If TR1 base voltage is +6·0V, its emitter voltage will be about +5·4V providing a forward bias of 0·6V.

If the signal strength now increases, the a.g.c. voltage increases and the voltage at point A will go more positive. This will increase the forward bias of TR2 stage causing its current to increase and its gain to reduce (forward gain control). As TR2 current increases so does the d.c. voltage across R1 which causes TR1 base potential to rise. This results in an increase in TR1 current and a reduction in its gain (forward gain control). The decrease in the gain of both TR1 and TR2 compensates for the rise in signal strength and thus tends to maintain a constant output signal from the receiver. The opposite effect will take place if the signal strength reduces causing the a.g.c. voltage at point A to become less positive and for the forward bias of TR1 and TR2 to reduce, which will result in an increase in their voltage gains.

## QUESTIONS ON CHAPTER FOUR

(1) The overall voltage gain in dBs of the system shown below in Fig.4.38 will be:

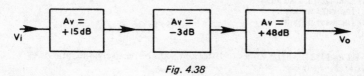

*Fig. 4.38*

    (a) +717dB
    (b) −717dB
    (c) +60dB
    (d) +66dB.

(2) If the input signal voltage to the system shown in Fig.4.38 above is 1mV, the output voltage will be about:
    (a) 1V
    (b) 100mV
    (c) 200mV
    (d) 6mV.

(3) An R–C coupled amplifier will amplify:
    (a) A.C. signals only
    (b) A.C. and D.C. signals
    (c) D.C. signal only
    (d) Audio signals only.

(4) When the voltage gain of an amplifier has fallen to 0·707 of maximum gain, the response will be:
    (a) 6dB down
    (b) 3dB down
    (c) 7·07dB down
    (d) 1dB down.

Questions 5–7 refer to the circuit of Fig.4.5 on page 56.

(5) If the capacitor C3 becomes open circuit, the effect will be:
    (a) Distorted output signals
    (b) Only half-cycles will be obtained at the output
    (c) No output
    (d) Loss of high frequency response only.

(6) If R2 becomes open circuit the effect will be:
    (a) Distorted output signals
    (b) Change in TR2 d.c. potentials
    (c) Smaller current in TR1
    (d) No output.

(7) If R1 becomes open circuit the effect will be:
    (a) No output signals
    (b) Larger current in TR1
    (c) Very distorted output
    (d) High output signals.

Questions 8 and 9 refer to the circuit of Fig.4.17(a) on page 71.

(8) If the value of the resistor R1 is increased the effect will be:
    (a) Change in TR1 bias
    (b) Increase in n.f.b.
    (c) Increase in gain
    (d) Reduction in gain.

(9) If R1 and R2 have the values of 10k$\Omega$ and 500$\Omega$ respectively, the ratio of $V_f:V_o$ will be:
    (a) 1:21
    (b) 1:20
    (c) 500:1
    (d) 20:1

(10) To obtain a narrow bandwidth in a tuned amplifier one would use:
    (a) Resistance damping
    (b) Stagger tuning
    (c) A high Q circuit
    (d) An integrated circuit.

(11) Variable tuned amplifiers are used in:
    (a) I.F. amplifiers
    (b) Audio amplifiers
    (c) R.F. amplifiers
    (d) I.C. amplifiers.

(12) A cascode r.f. amplifier may consist of:
    (a) A common emitter and a common base amplifier in series
    (b) A common base and common collector amplifier in parallel
    (c) A common source and a common drain amplifier in series
    (d) A common base and a common gate amplifier in series.

(13) A ceramic filter has:
    (a) A high gain
    (b) Bandpass characteristics
    (c) An adjustable centre frequency
    (d) A low Q.

(14) The 3dB Bandwidth of an f.m. radio receiver i.f. amplifier will probably be about:
    (a) 10 kHz
    (b) 240 kHz
    (c) 100 MHz
    (d) 5·5 MHz.

(15) The effect on a transistor to which forward gain control is applied when the signal strength increases is:
    (a) Reduction in forward bias
    (b) Increase in gain
    (c) Increase in forward bias
    (d) Decrease in collector current.

# CHAPTER FIVE

# POWER AMPLIFIERS

POWER AMPLIFIERS have to provide power gain as opposed to voltage amplifiers where the need is for voltage gain. Since power is equal to $V \times I$, a power amplifier must provide appreciable voltage or current gain or a combination of both.

### Need for Power Amplification

When transducers, particularly the large electrical-to-mechanical types, are used in electronics appreciable work must be done on them for satisfactory operation, thus appreciable electrical power must be supplied to them. Typical examples are a loudspeaker in an audio amplifier, a motor in a servo control system or an ultrasonic vibrator used in a flaw-detection system. If the signal source for the transducer is of low power, a power amplifier is required between the signal source and the transducer. The transducer therefore constitutes the load for the power amplifier.

### Class-A and Class-B Power Amplification

Class-A operation of a power amplifying transistor is illustrated in Fig.5.1(a). With this class of operation, the transistor is provided with a forward bias such that with the given sinewave input, the collector current is never cut off. Operation is confined to the linear part of the characteristic, i.e. between A and C to keep non-linear distortion to a minimum. In the absence of signal drive the quiescent collector current is fairly high. If when drive is applied the output current is symmetrical on both half-cycles, the mean current is the same as the quiescent current.

Class-B operation is illustrated in Fig.5.1(b). In power amplifiers using this method, the transistor is biased so that with sinewave drive the collector current flows for half the cycle. Thus in true class-B the forward bias is zero and the collector current in the absence of signal drive is zero. The mean current, which is less than in class-A, increases with the drive signal amplitude. Because of the need for a small forward bias to produce current flow and the

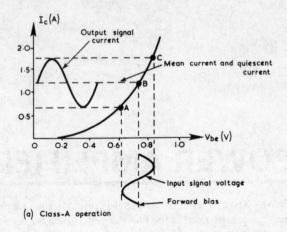

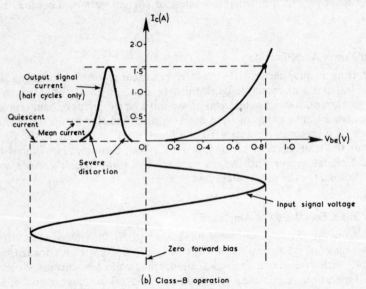

Fig. 5.1 Class-A and Class-B operation in power amplifiers.

initial non-linearity of the characteristic, severe distortion of the half-cycles will occur as shown. For resistive or non-resonant loads as with an audio amplifier, class-B can only be used in push-pull output stages.

### Class-A Power Amplifier

A class-A power amplifier stage is shown in Fig.5.2. TR1 is a power type transistor and is usually fitted with a heat sink. The transistor which may be n–p–n or p–n–p is biased in the usual way by the potential divider R1, R2 to class-A operation. Resistor R3 provides d.c. stabilisation of the operating point and is suitably decoupled by C2 to prevent n.f.b. The load for the amplifier is represented by the resistor $R_L$ which for an audio amplifier would be a loudspeaker. Maximum power will be delivered to a load when the load resistance is equal to the internal resistance of the signal source. In an audio amplifier, maximum power transfer is not the only consideration and some attention must be given to the amount of distortion present. Thus it is normal to match the load resistance to the optimum load resistance of the output transistor. The optimum load is one which gives good power transfer consistent with minimum distortion. When the actual load resistance is less than the optimum load, a transformer may be used for matching. This is the purpose of the output transformer T2 which uses a step-down ratio. The use of a transformer eliminates the d.c. component from the load, which in the case of a loudspeaker would cause the cone to be off centre. The load seen by TR1 is equal to $n^2 R_L$.

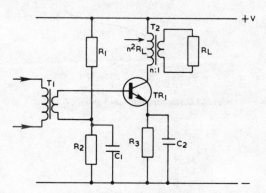

*Fig. 5.2 Class-A power amplifier.*

The signal to receive power amplification by TR1 may be coupled from the previous stage (called the driver) by a transformer T1 as shown, or may be d.c. coupled. The use of a driver transformer permits correct matching between the output resistance of the driver stage and the input resistance of TR1. Normally T1 will utilise a step-down ratio. Class-A power amplifiers using a single transistor are not common but are sometimes used in the audio stages of a car radio or television receiver where the high quiescent current is not of major importance.

Although with class-A working, operation is said to be confined to the linear part of the transistor characteristic, there is always some non-linearity present which can give rise to distortion. Consider Fig.5.3 where point B is the quiescent operating point. If a large signal is applied so that the operation is between the limits of points A and C, the output current will be distorted as shown due to the non-linearity of the characteristic. There would be a

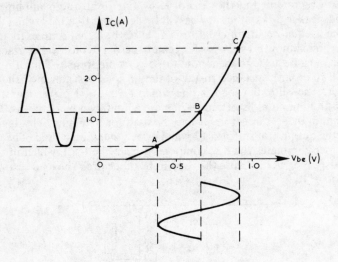

*Fig. 5.3 Distortion in class-A power amplifier.*

greater distortion at high power levels than lower power levels. Since with power amplifiers we are concerned with larger signal levels than in a voltage amplifier, the power stage of an amplifier will be a key factor in determining the overall quality. Distortion of this nature can be reduced by applying a.c. negative feedback over the output stage. At the same time any n.f.b. used will assist in improving the frequency response. In Fig.5.2 the frequency response is limited at low frequencies by the inductance of the transformers T1 and T2 and at high frequencies by the transistor and/or the transformers.

The bandwidth of a power amplifier is the band of frequencies between the limits where the power gain has fallen to half its mid-band power level, see Fig.5.4. In decibels the half-power points are 3 dB down. The signal voltage across the load will be 0·707 of the mid-band voltage at the half-power points.

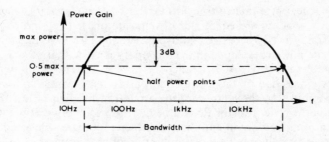

Fig. 5.4 Bandwidth of power amplifier.

## Class-A Push-pull Power Amplifier

When more power is required than can be given by a single class-A power amplifier stage, push-pull operation is invariably used. If the main requirement is low distortion, class-A push-pull is used.

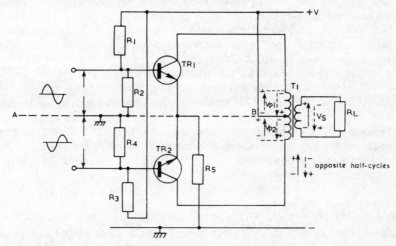

Fig. 5.5 Basic class-A push-pull power amplifier.

A basic circuit is given in Fig.5.5. Two power transistors TR1 and TR2 are required which are biased to class-A by their potential dividers: R1, R2 for TR1 and R3, R4 for TR2. The output currents of the transistors are combined by a centre-tapped transformer T1 to feed a common load $R_L$. With push-pull, equal but opposite phase drive signals are required as shown. When applied to the bases, the drive signals affect the transistors in opposite senses, i.e. as TR1 current increases on a positive half-cycle of its input, TR2 current decreases as its input will be on a negative half-cycle. The opposite effect occurs on the following half-cycle when the input signal changes polarity. Thus TR1 and TR2 collector currents are also equal and of opposite phase.

Because of the centre-tapped transformer the a.c. voltages across each half of the primary winding are equal and of opposite phase ($V_{p1}$ and $V_{p2}$). These primary voltages are therefore additive in developing the secondary voltage $V_s$. Thus the power developed in the secondary is twice that of a single transistor operating in class-A. In reality we have two identical class-A power amplifiers either side of the dotted line AB driven in antiphase with their outputs combined by the centre-tapped output transformer.

A common-emitter resistor R5 is used but this need not be decoupled since the a.c. currents in it are in antiphase and therefore there will be no n.f.b. at a.c. The resistor will provide d.c. stabilisation of both transistors in the usual way. Normally, the transistors will have closely matched characteristics, i.e. a matched pair will be used.

Class-A push-pull has the following advantages:

(1) The d.c. collector currents of TR1 and TR2 magnetise the core of the output transformer in opposite directions. Hence, with equal collector currents there is no d.c. magnetisation of the core. The output transformer can therefore be made smaller than for an equivalent class-A single transistor power stage.

(2) Because TR1 and TR2 collector currents are equal and of opposite phase, the a.c. current flowing in the internal resistance of the power supply is zero, thus supply line decoupling is simplified.

(3) Any currents in the primary of the output transformer due to 50 Hz or 100 Hz ripple on the d.c. supply cause magnetisation in opposite directions in the core. This results in less hum in the load compared with a single class-A stage.

(4) Because of the push-pull operation any even harmonic distortion produced by the output transistors is eliminated. Thus for the same amount of distortion, the power output may be more than twice that of a single class-A stage.

### Phase Splitting Circuits

Circuits used to provide equal but opposite phase drive signals to a push-pull amplifier are called phase-splitters. A common method is to use a centre-tapped driver transformer as shown in Fig.5.6(a).

The signal input voltage applied to the primary will produce voltages in each half of the tapped secondary as shown (represented by the arrows). On each half-cycle of the input, the voltages at either end of the secondary will be in antiphase with respect to the centre tap which is connected to a neutral point.

A transistor may be used as a phase-splitter and one example is shown in diagram (b). Here outputs are taken from the collector and emitter of the transistor which is supplied with a signal at its base. Since the collector signal voltage is in antiphase with the base voltage, and the emitter signal voltage is in phase with the base, the output signals across R1 and R2 will be in

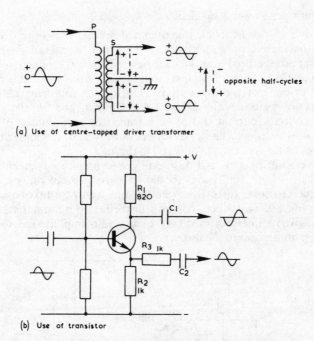

(a) Use of centre-tapped driver transformer

(b) Use of transistor

*Fig. 5.6 Phase-splitters.*

antiphase with each other. Resistor R3 is added to increase the output re-sistance of the emitter to match that of the collector. As there will be some signal loss across R3, the emitter load R2 is made larger than the collector load R1.

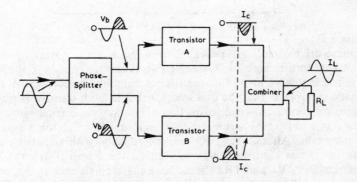

*Fig. 5.7 Concept of two transistors in class-B push-pull.*

## Class-B Push-pull Power Amplifiers

Power output transistors may be operated in class-B push-pull and Fig.5.7 shows the basic concept. As with class-A push-pull, equal but antiphase drive signals ($V_b$) are required and they are derived from the phase-splitter. With class-B operation transistor A is, say, cut off whilst transistor B is conducting for one half-cycle of the input signal. During the following half-cycle the conditions are reversed with transistor A conducting and transistor B cut off. Thus the collector current ($I_C$) of each transistor is composed of half-cycle current pulses as shown. These half-cycles must then be combined to produce a continuous sinewave current flow ($I_L$) in the load.

A basic circuit for a class-B push-pull power amplifier is given in Fig.5.8. In true class-B both transistors should be biased to cut off, but with zero forward bias crossover distortion occurs as one transistor takes over from the other, see Fig.5.9. To reduce crossover distortion to a minimum, both transistors are given a small forward bias. In an audio amplifier crossover distortion is most unpleasant to the listener.

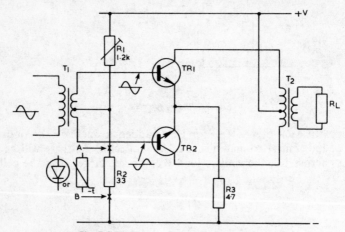

*Fig. 5.8 Basic class-B push-pull output stage.*

Antiphase drive signals are provided from the secondary winding of the phase-splitting transformer T1. A small forward bias is supplied to the bases via the secondary winding of T1 from the potential divider R1, R2. Resistor R1 is made variable so that the quiescent current in both transistors may be set to give minimum crossover distortion. This may be carried out by inserting a d.c. ammeter in series with R3 and adjusting R1 for a specified quiescent current. Alternatively, R1 may be adjusted with signal input for minimum crossover distortion, as observed on the screen of a c.r.o. connected across $R_L$. TR1 and TR2 should have matched characteristics, otherwise distortion will arise due to inequality in the two half-cycles of collector current. The half-cycles are combined in the load $R_L$ because of the use of the

centre-tapped output transformer T2. A common emitter resistor R3 is used
which must not be decoupled since the current in R3 is composed of uni-
directional half sinewaves. A decoupling capacitor would therefore build up
a d.c. voltage across it which would tend to reverse bias the transistors.

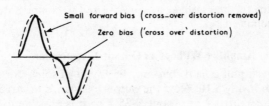

Fig. 5.9 Need for small forward bias in class-B.

The main advantage that class-B has over class-A is the low quiescent current and higher efficiency at full output. Because of the low quiescent current,
class-B is always chosen for power amplifiers operating from dry batteries.

### Bias Stabilisation

The small forward bias provided in class-B is rather critical. If it is too
small crossover distortion will occur and if too large the quiescent current is
increased. The operation of the basic circuit would be satisfactory for a fixed
supply line voltage and a fixed temperature. However, if the temperature or
the supply line voltage alters the circuit would be unsatisfactory for the following reasons.
(1) As the temperature rises, the base-emitter voltage drop decreases for a
    given emitter current (an effect that occurs in any p-n junction as ex-
    plained in Volume 2). Thus, if the bias components R1 and R2 are chosen
    to give the correct bias at low temperature, the quiescent current of both
    transistors will rise as the temperature increases. The increase in quiescent
    current is an important factor when the amplifier is operated from a dry
    battery.
(2) The bias voltage across R2 is dependent upon the supply voltage. If the
    supply voltage falls (assuming dry battery supply), the bias will fall result-
    ing in crossover distortion.
The common-emitter resistor R3 will provide a measure of d.c. current
stabilisation, but its value is usually limited because part of the output power
of the stage is lost in R3 and the d.c. voltage drop across it subtracts from
the available collector-to-emitter voltage. Its value is therefore quite small
and its contribution to d.c. stabilisation consequently lowered.
To help overcome these problems either an n.t.c. thermistor or a p-n diode
may be fitted in place of R2. The p-n diode idea is the most satisfactory
method because if the supply line voltage alters, the voltage drop across the
diode remains constant thus the bias is constant. Also, if the temperature

increases, the voltage drop across the diode decreases which compensates for the decrease in voltage drop across the base-emitter junctions of the output transistors (assuming the diode and transistors are the same type, i.e. silicon). A thermistor on the other hand can only compensate for temperature variations, e.g. with a rise in temperature the resistance of an n.t.c. type will decrease causing the bias voltage to decrease and so compensate for the lower base-emitter voltage drop.

### Push-pull Power Amplifier Without an Output Transformer

A class-B push-pull circuit that may be found in some portable radio receivers is given in Fig.5.10. Now the output resistance of large and medium power transistors varies from about 5–75$\Omega$, so it is possible to dispense with the output transformer and match directly between the output transistor and a loudspeaker load of suitable impedance. However, if the stage is to be used in push-pull the output currents from the two transistors must be directed so that they combine in the load.

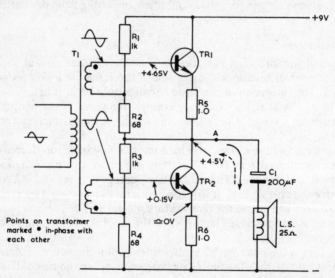

*Fig. 5.10 Class-B push-pull eliminating need for output transformer.*

As far as d.c. is concerned TR1 and TR2 are in series across the d.c. supply and the d.c. potential at point A is half the supply potential. Because of the series connection the bases of the transistors cannot be supplied with the same d.c. voltage as was used in Fig.5.8. Thus the antiphase drive signals have to be supplied from separate windings on the driver transformer T1. To prevent crossover distortion a small forward bias is supplied to both transistors; R1, R2 provide the bias for TR1 and R3, R4 provide the bias for TR2. Resistors R5 and R6 are for d.c. stabilisation.

For alternating currents the two transistors are in parallel and are driven in antiphase by the signals from T1 secondaries. Let us say that the signal applied to TR1 base is on a positive half-cycle, therefore the signal applied to TR2 base will be on a negative half-cycle. TR1 will thus conduct hard and TR2 will cut off (class-B operation). The potential at point A will now rise and current will flow via C1 to the loudspeaker in the direction shown by the solid line. Conversely, on the following half-cycle of the input when TR1 base is taken negative and TR2 base goes positive, TR1 cuts off and TR2 conducts hard. As a result the potential at point A falls and current flows in the opposite through the loudspeaker (and C1) in the direction shown by the dotted line. Thus TR1 supplies the current to the loudspeaker during one half-cycle and TR2 supplies the current during the other half-cycle. C1 blocks the d.c. from the loudspeaker and passes only the changes in potential at point A to the load. The value of C1 must be such that it offers a low reactance path to the signal currents at low frequencies. Typical voltages are given for germanium transistors. Although n-p-n transistors are shown, p-n-p transistors may be used with a reversal of the supply line potential. The circuit will operate in class-A if required but is not normally used in battery operated equipment.

An advantage of this circuit arrangement is that an output transformer is not required, which permits a more economical design. The loudspeaker may alternatively be returned to the positive supply line without affecting the principle of operation.

## Complementary Symmetry Power Amplifiers

With push-pull operation the current in one transistor increases whilst the current in the other decreases. If the transistors are of the same type (two n-p-n or two p-n-p) this is achieved by driving them in antiphase. The same action can be obtained by using complementary transistors, i.e. one p-n-p and one n-p-n and driving them in phase.

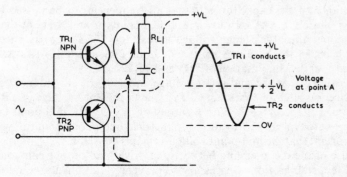

Fig. 5.11 Basic form of complementary symmetry circuit.

A matched pair of n-p-n and p-n-p power transistors are arranged in the basic circuit shown in Fig.5.11 to form a complementary symmetry circuit. Because of the simpler in phase drive for this circuit, a phase-splitter is not required. Also, the output transformer may be dispensed with since the transistors may be matched to the load as for the previous circuit.

For d.c. the two transistors are arranged in series thus the d.c. potential at point A is half the line supply potential. With no signals applied, the capacitor C will be charged to $0.5V_L$. Assuming true class-B operation for the present, when the input signal is on a positive half-cycle TR1 will be biased on and TR2 biased off. Conversely, when the input signal is on a negative half-cycle, TR1 will be biased off and TR2 biased on.

If we consider maximum drive conditions when TR1 conducts hard and TR2 is off, current is supplied to the load via the conducting TR1 and flows in the direction shown by the solid line. This causes C to discharge to zero and for the voltage at point A to rise up to a voltage equal to $V_L$. On the other half-cycle when TR1 is off and TR2 is conducting hard, current is supplied to the load via the conducting TR2 and flows in the direction shown by the dotted line. This causes C to charge up to the full supply voltage $V_L$ and for the voltage at point A to fall to zero. It will be seen that the action permits the current to reverse direction in the load every half-cycle as is required.

### Practical Circuit

A practical arrangment for a class-B circuit suitable for use as an audio power output stage is given in Fig.5.12. The complementary output pair is formed by TR1 and TR2 with TR3 operating as a driver stage. The bases of TR1 and TR2 are driven in phase by the signal from across the collector load (R1) of the driver transistor TR3. Resistor R2 applies a small forward bias to TR1 and TR2 to overcome the objectionable crossover distortion. The value of this resistor is chosen so that TR1 base voltage is slightly positive w.r.t. its emitter whilst TR2 base is slightly negative w.r.t. its emitter. Capacitor C1 blocks the d.c. from the mid-point A; a large value is chosen so that the reactance of the capacitor is low at all frequencies to be passed by the amplifier. Since C1 is of very low reactance, points A and B are at the same a.c. potential thus the signal voltage across R1 (TR3 load) is effectively applied between the base and emitter of both TR1 and TR2. Resistors R3 and R4 are for d.c. stabilisation of TR1 and TR2.

It was stated that the mid-point voltage was half the supply line voltage. In practice it is slightly offset from $0.5V_L$ due to the knee voltage of TR1 and the emitter voltage of TR3. With a supply of, say, 9V the mid-point voltage would be set to typically 5V to allow symmetrical operation about this voltage. It is important that the mid-point voltage remains constant.

Because of the d.c. coupling between the driver and output pair some additional d.c. stabilisation is necessary to prevent variations in the mid-point voltage. This is done by deriving the base bias for the driver from the mid-

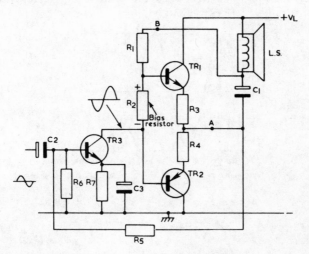

*Fig. 5.12 Practical arrangement for complementary symmetry push-pull power amplifier (class-B).*

point voltage via R5, forming d.c. negative feedback. For example, suppose that the mid-point d.c. potential rises slightly due to a variation in TR3 current. This will cause the base bias on TR3 to increase and for its collector current to rise. As a result there will be a larger voltage drop across both R1 and R2 which will cause TR1 to conduct less and TR2 to conduct more. This will cause the mid-point voltage to fall, thereby compensating to some extent for the original rise. In addition to the d.c. feedback via R5 there will be a.c. n.f.b. which will assist in reducing distortion introduced by the driver and output pair and improve the frequency response.

### Transistor Dissipation

Something will now be said about transistor dissipation as it is rather important. Fig.5.13(a) shows the voltage at the mid-point and the current in the on transistor under maximum power output conditions. Now this condition does not result in maximum transistor dissipation as at first might be thought. Although the current in the on transistor is at its greatest, the voltage across the transistor is very small (almost zero when the current is at a peak). Maximum transistor dissipation occurs at a lower level of output power when the peak voltage at the mid-point (or across the load) is equal to 0·63 of 0·5$V_L$, see diagram (b).

This is also shown in Fig.5.14. Curve A is the power taken from the d.c. supply and curve B is the power developed in the load. The difference between curve A and curve B is the power dissipated in the transistor, curve C. For example, if a d.c. supply of 10V is used for the output pair, maximum transistor dissipation will occur when the peak voltage across the load is $0·63 \times 0·5 \times 10$ Volts $= 3·15V \, (\simeq 3V)$.

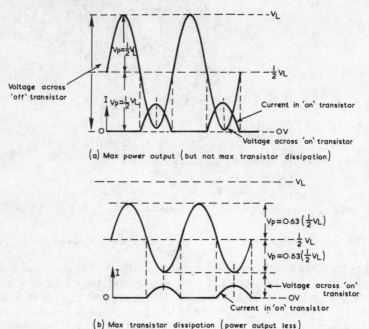

(a) Max power output (but not max transistor dissipation)

(b) Max transistor dissipation (power output less)

*Fig. 5.13 Transistor dissipation in class-B.*

### Effect of Load Resistance on Transistor Dissipation

(a) If the load resistance is doubled, the power supplied, power in load and transistor dissipation are all halved.

(b) On the other hand if the load resistance is halved, the power supplied, power in load and transistor dissipation are all doubled.

It will be appreciated that (b) is very important because if a class-B amplifier has been designed to supply, say, a 15-ohm load but is operated with a 7·5-ohm load, the amplifier is liable to be damaged due to excessive transistor dissipation. If the load is accidentally short-circuited, transistor dissipation will be even higher.

### Setting the Mid-point Voltage

Some means of setting the mid-point voltage of the output pair may be included in the circuit and an example is shown in Fig.5.15. An a.f. amplifier TR1 is used prior to the driver stage with d.c. coupling throughout from TR1 up to the output pair. The preset resistor R2 is included to allow the mid-point voltage to be accurately set. Adjustment of R2 alters TR1 current so altering

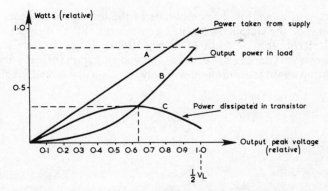

*Fig. 5.14 Relationship between peak voltage across load and supply power, output power and transistor dissipation.*

its collector voltage, the base and collector voltages of TR2, the base voltages of TR3 and TR4 and finally the mid-point voltage.

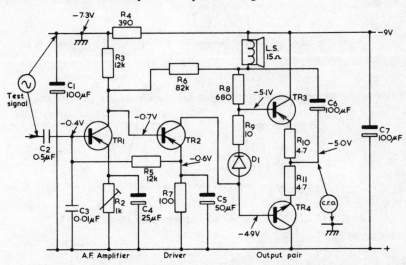

*Fig. 5.15 Adjustment of mid-point voltage for class-B complementary symmetry output stage.*

There are two ways of setting the mid-point voltage:

(1) Under Quiescent Conditions

With a d.c. voltmeter connected between the junction of R10, R11 and the positive supply line, R2 is set for a nominal mid-point voltage of 5V (for this circuit).

(2) Under Dynamic Conditions

A more accurate method is to apply a test signal of, say, 1 kHz between the input to TR1 and chassis and to connect a c.r.o. between the mid-point and chassis as shown. The amplitude of the test signal is increased until clipping of the output signal just commences. R2 is then set for symmetrical clipping as indicated in Fig.5.16(a). If R2 is set incorrectly, asymmetrical clipping will occur as shown in diagram (b), i.e. symmetrical operation about the mid-point will only occur on low level drive signals.

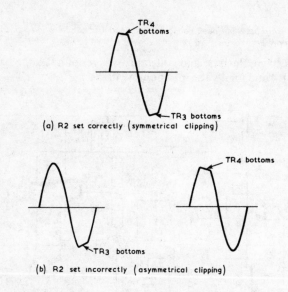

Fig. 5.16 Waveforms obtained at mid-point during adjustment of $R_2$.

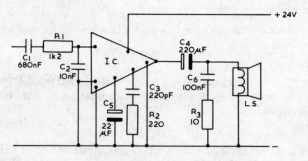

Fig. 5.17 I.C. Class-B audio power amplifier.

### I.C. Power Amplifier

Power amplifiers are available in integrated circuit form, but the power output is somewhat limited to, say, about 5–10W. An example is given in Fig.5.17 where the i.c. is included with a number of external discrete components. The transistors formed within the i.c. are often arranged in a class-B complementary symmetry circuit with the i.c. attached to a heat sink to prevent high temperatures inside the i.c. Provision may be made for external feedback to be connected to reduce distortion and improve the frequency response. The input signal to the i.c. is applied via C1 and R1 and the output from the i.c. applied via the coupling capacitor C4 to the loudspeaker load.

## QUESTIONS ON CHAPTER FIVE

(1) With class-B operation of a transistor, collector current flows for:
(a) 90°
(b) 180°
(c) 270°
(d) 360°
of the input cycle.

(2) At the half-power points of a power amplifier, the voltage across the load will be:
(a) 6dB up
(b) 6dB down
(c) 3dB up
(d) 3dB down.

(3) The voltages at either end of a tapped transformer winding are w.r.t. the tap:
(a) In phase
(b) 90° out of phase
(c) In antiphase
(d) 270° out of phase.

(4) If the load ($R_L$) for the amplifier shown in Fig.5.2 on page 95 is $5\Omega$ and T2 uses a 3:1 step-down ratio, the effective load for TR1 will be:
(a) $5\Omega$
(b) $15\Omega$
(c) $30\Omega$
(d) $45\Omega$.

(5) An audio power amplifier using a single transistor may work in:
(a) Class-B only
(b) Class-A and class-B
(c) Class-A, -B or -C
(d) Class-A only.

(6) If R1 goes open circuit in Fig.5.8 on page 100, the effect will be:
(a) No output
(b) Crossover distortion in the output
(c) Half-cycles only in the output
(d) Excessive transistor dissipation.

(7) If the mid-point voltage of Fig.5.12 on page 105 falls, the effect will be:
(a) An increase in TR3 collector current
(b) An increase in TR1 and TR2 currents
(c) An increase in TR2 current and a decrease in TR1 current
(d) An increase in TR1 current and a decrease in TR2 current.

The following questions refer to Fig.5.15 on page 107
(8) The bias for the output pair is provided by:
(a) R10 and R11
(b) R8
(c) R9
(d) R9 and D1.

(9) The purpose of D1 is:
   (a) Stabilise the current in TR2
   (b) Stabilise the current in TR3 and TR4
   (c) Reduce the n.f.b. in TR3
   (d) Reduce the n.f.b. in TR4.

(10) If the 15Ω loudspeaker is replaced by one of 5Ω the effect will be:
   (a) Less output power
   (b) Less power taken from d.c. supply
   (c) Higher dissipation in TR3 and TR4
   (d) Lower dissipation in TR3 and TR4.

(11) Maximum power will be developed in the load when the peak voltage at the mid-point is:
   (a) About 8V
   (b) About 8·9V
   (c) About 4V
   (d) Exactly 9V.

(12) If R5 goes open circuit the effect on TR1 current will be:
   (a) No change
   (b) Slight increase
   (c) Large increase
   (d) Fall to zero.

(13) Resistor R6 provides:
   (a) A.C. negative feedback
   (b) D.C. positive feedback
   (c) D.C. stabilisation of the mid-point voltage
   (d) D.C. coupling between TR1 and the output pair to reduce drift.

(14) The emitter voltage of TR1 (germanium transistor) would be typically:
   (a) −0·4V
   (b) −1·0V
   (c) −0·8V
   (d) −0·25V.

(15) D.C. negative feedback is provided by:
   (a) R8
   (b) R5
   (c) R9
   (d) R3.

(16) The purpose of C7 is to:
   (a) Lower the internal resistance of the d.c. supply to a.c.
   (b) Allow a high value of load for the output pair to be used
   (c) Provide a small amount of overall positive feedback
   (d) Prevent excessive power dissipation in the output pair.

# CHAPTER SIX

# SINEWAVE OSCILLATORS

WHEN DISCUSSING AMPLIFIERS in Chapter 4 it was stated that positive feedback is to be avoided otherwise instability will arise. In some circuits, positive feedback is deliberately introduced, i.e. a fraction of the output signal is fed back in phase to the input. An amplifier with positive feedback will provide a gain of:

$$A' = \frac{A}{1 - AB}$$

where A is the gain without feedback and B is the fraction of the output that is fed back. As an example consider an amplifier with a voltage gain of 100 without feedback and with 1/100 of its output fed back to the input in phase. The gain with feedback will be:

$$A' = \frac{100}{1 - (100 \times \frac{1}{100})}$$

$$= \frac{100}{1 - 1}$$

$$= \frac{100}{0}$$

$$= \infty$$

Thus the gain with positive feedback is now infinite which means that the amplifier is providing its own input and no longer operates as an amplifier to external signals. It is now acting as an oscillator and will provide its own input and produce a continuous output independent of any external signal input. The term AB is known as the loop gain and if this is equal to or greater than unity, the gain with feedback becomes infinite and oscillation will occur. In practice the gain cannot be infinite as the signal output will become limited and thus limit the gain.

## L.C. Sinewave Oscillators

Oscillators that give a continuous sinewave output in the frequency range of 30 kHz to 200 MHz or so, use a frequency determining circuit comprising inductance and capacitance. These are called L.C. oscillators and Fig.6.1 shows the basic form of all such oscillators.

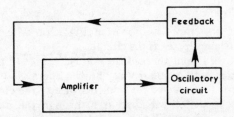

Fig. 6.1 Basic form of all L.C. oscillators.

An L.C. oscillator can be arranged into three parts:
   (a) A frequency determining L.C. network (the oscillatory circuit)
   (b) A feedback network
   (c) An amplifying device.
The oscillatory circuit generates the sinewave oscillation and a fraction of its output is fed back via the feedback network to the input of the amplifier. After amplification the feedback signal is reapplied to the oscillatory circuit.

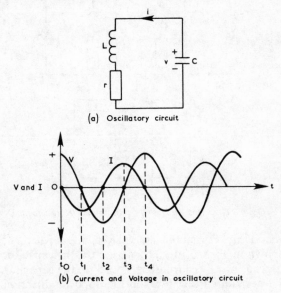

(a)  Oscillatory  circuit

(b) Current and Voltage in oscillatory circuit

Fig. 6.2 The L.C. oscillatory circuit.

If the gain of the amplifier is A and the fraction fed back is B, the loop gain is AB. For oscillation to occur, the loop gain must be equal to or greater than unity and the feedback signal must be reapplied in phase to the oscillatory circuit.

### The Oscillatory Circuit

An L.C. oscillatory circuit is shown in Fig.6.2(a) where the resistance r represents the resistance losses of the circuit.

Consider that C has been charged from a d.c. source to produce a voltage v between its plates with a polarity as shown. Energy is then stored in the capacitor in the form of an electric field. Because of the circuit produced by the inductor, the capacitor will discharge through the inductor causing a current i to flow as indicated. If the current commences to flow at instant $t_o$, see diagram (b), it will reach a maximum at instant $t_1$ at which time the capacitor will be fully discharged. The energy that was initially stored in C is now stored in L in the form of a magnetic field. As there is now no voltage to support the current in L, the magnetic field will commence to collapse. In doing so, an e.m.f. will be induced in L with a polarity as shown in Fig.6.3(d). The direction of the induced e.m.f. is such that it keeps the current flowing in the same direction.

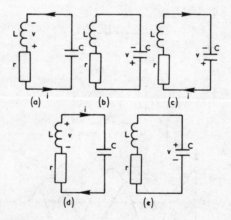

Fig. 6.3 Diagrams explaining action in oscillatory circuit.

As the current decays during the interval $t_1$–$t_2$, the capacitor recharges with the polarity shown in Fig.6.3(b) and at instant $t_2$ the capacitor will be fully charged. Energy has now been transferred back to the capacitor. The capacitor now discharges and current flows in the opposite direction through the inductor as in diagram (c). This occurs during the time interval $t_2$–$t_3$. At instant $t_3$, the capacitor will be fully discharged and the current in L at a maximum. As there is no voltage to support current flow, the magnetic field of L will collapse inducing an e.m.f. as shown in diagram (d). The current continues to flow in the same direction but decays during the interval $t_3$–$t_4$ as C recharges. At instant $t_4$, the capacitor will be fully charged with a polarity the same as it started with and the current will be zero, diagram (e). This completes one cycle of events and the process may be repeated.

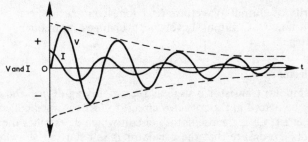

Fig. 6.4 Damped oscillation.

It will be seen that there will be a continuous exchange of energy between L and C. Because of the presence of the circuit resistive losses r, each time current flows there will be an energy loss. Thus, unless this loss can be made up in some way, the voltage and current waveforms will gradually decay as shown in Fig.6.4 until they are zero, when the oscillation will stop. This is called a damped oscillation.

### Continuous Oscillation

If energy can be supplied to the LC circuit to replace the energy lost in r,

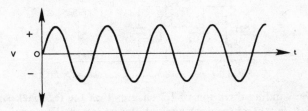

Fig. 6.5 Continuous oscillation.

a continuous or undamped oscillation can be maintained as shown in Fig.6.5. When the oscillatory circuit losses are small, the frequency of the oscillation is given by

$$f_o = \frac{1}{2\pi \sqrt{L.C}}$$ L in henrys and C in farads.

### L.C. Oscillator Circuits

A large number of different oscillator circuits are possible. Some are intended for fixed frequency operation and others for variable frequency operation, e.g. in test signal generators and local oscillators in receivers. In arriving at a suitable circuit, the following factors are given particular attention:

    (a) Stability of Frequency (with temperature, supply voltage and load variations)

    (b) Purity of Output Waveform (freedom from harmonics)

    (c) Constancy of Output Level (with changes in frequency or supply voltage).

### Tuned Collector

When a bipolar transistor is used as the amplifying device, the oscillatory circuit may be placed in the collector circuit to form a tuned collector circuit as in Fig.6.6. L1, C1 is the oscillatory circuit which determines the frequency of operation. To ensure that the oscillator is self-starting at switch-on, the transistor is given a small forward bias from the potential divider R1, R2. Feedback to the base of the transistor is via a small coupling winding L2 which is inductively coupled to L1. The turns ratio and the degree of coupling determine the amount of signal that is fed back to the base.

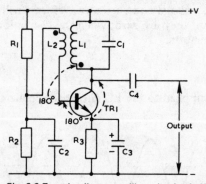

Fig. 6.6 Tuned collector oscillator (series fed.)

Also, the winding direction of L2 ensures that the feedback signal to the base is 180° out of phase with that at the collector. The oscillatory signal applied to the base produces a signal at the collector 180° out of phase with that at the base. Thus the signal arriving back at the oscillatory circuit is in phase as is required to sustain the oscillation.

Normally, the amount of feedback provided by L2 is greater than is required to make good the losses in L1, C1, i.e. the loop gain is greater than unity. Once oscillation commences, it will tend to build up in amplitude, so some means must be provided to limit the amplitude. Note that it would be

very difficult to maintain exactly the right amount of feedback by means of
L2; if it were to fall by a small amount the oscillation would cease even though
it may have started.

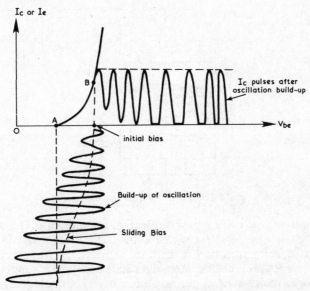

Fig. 6.7 Operation of oscillator showing action of sliding bias.

In this circuit, the magnitude of the oscillation is limited by the action of
R3, C3 which may be explained with the aid of Fig.6.7. Suppose that the
initial bias from R1, R2 biases the transistor to point B. After a few cycles as
the oscillation builds up to a large amplitude, collector and emitter current
will flow only during positive half-cycles. This causes C3 to become charged
with polarity as shown, thereby reducing the effective base emitter forward
bias. The greater the amplitude of oscillation, the larger the voltage across C3
and hence the smaller is the forward bias. As the forward bias reduces, the
period for which the transistor is conducting is shortened and the energy fed
to L1, C1 is less. As equilibrium is reached where the energy fed to L1, C1 is
just sufficient to make up for the losses, the amplitude of oscillation remains
constant. At this point the loop gain will become unity. If any change takes
place, e.g. the supply voltage alters, the voltage across C3 will alter to re-
establish equilibrium conditions.

The gradual decrease in bias of the transistor is sometimes called sliding
bias. It will be seen that once the oscillation has reached equilibrium, the bias
on the transistor corresponds to point A, i.e. class-B (in some circuits it may
go into class-C). The time constant of R3, C3 is important because during
the negative half-cycles when the transistor is non conducting, C3 discharges
through R3 and the bias should not change appreciably during this period.

The value of C3 should not be too large or the bias on the transistor will not follow changes in the amplitude of oscillation. Capacitor C2 decouples R2 to signal to prevent loss of feedback signal across R2.

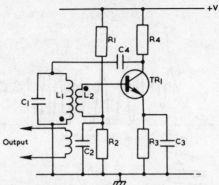

*Fig. 6.8 Tuned collector oscillator (shunt fed).*

An alternative tuned collector oscillator is shown in Fig.6.8. The oscillatory circuit L1, C1 is now in shunt with the transistor and is fed via C4 (d.c. block). This capacitor will have a small reactance at the frequency of oscillation. Resistor R4 prevents the internal resistance of the d.c. supply damping the oscillatory circuit; an r.f. choke may alternatively be used for this purpose. Initial forward bias is provided by R1, R2 with C3, R3 producing the sliding bias action as for the previous circuit. The circuit action is identical with the series-fed arrangement, with the coupling winding L2 providing the antiphase feedback to the base. After inversion by the transistor the oscillatory signal is fed back in phase to L1, C1 via C4. An advantage of the shunt fed circuit is that when C1 is to be made variable, one side of it may be connected to the common or chassis line.

**Tuned Base**

Instead of placing the oscillatory circuit in the collector it may be connected in the base circuit of the transistor to form a tuned base oscillator as shown in Fig.6.9. The operation is generally similar, with L1, C1 forming the frequency determining circuit and L2 the feedback winding. The oscillatory signal applied to the base receives phase inversion by the transistor. After passing through C4 the signal receives further phase inversion by the coupling between L2 and L1 to arrive back in phase across L1, C1.

Capacitor C4 blocks the d.c. from the collector so that it is not shorted out via L2. As with the shunt fed tuned collector circuit, the tuned base arrangement permits one side of C1 to be connected to the common line. L1 may be tapped as shown to prevent the low input resistance of the transistor damping the oscillatory circuit excessively.

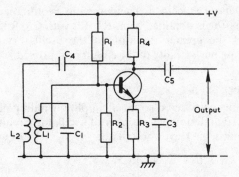

*Fig. 6.9 Tuned base oscillator.*

### Reinartz Oscillator

An oscillator that is commonly used as the local oscillator in a.m. radio receivers is shown in Fig.6.10. This circuit is connected in common base with C2 grounding the base to oscillatory signals.

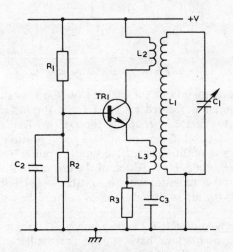

*Fig. 6.10 Reinartz oscillator.*

With the common base configuration, it is necessary for feedback through the transistor to be between emitter and collector circuits. The oscillatory circuit is formed by L1, C1 with C1 made variable for receiver tuning purposes. Once the oscillation has started the oscillatory signal is coupled from L1, C1 to the emitter via the coupling winding L3. After amplification by the transistor the signal is fed back to L1, C1 via another coupling winding L2. The winding directions of the coupling windings ensure that the energy fed back to the oscillatory circuit meets the usual in phase requirement. Initial starting bias for the transistor is obtained from R1, R2 with C3, R3 providing sliding bias. Because of the greater isolation of the oscillatory circuit from the transistor the frequency stability of the circuit is improved.

### Hartley Oscillator

In all the circuits considered so far a coupling winding has been used, but this is not essential and among a number of oscillator circuits that do not use a coupling winding is the Hartley oscillator.

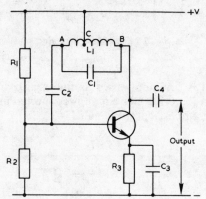

*Fig. 6.11 Hartley oscillator (series fed).*

A commonly used version of the Hartley oscillator is shown in Fig.6.11. The oscillatory circuit is formed by L1, C1 but the inductor is tapped to achieve the necessary phase inversion in the external feedback circuit. L1 is a continuous winding and if, say, point A is going positive w.r.t. the tap C, point B will be going negative w.r.t. the tap, i.e. points A and B will be in antiphase with each other. The position of the tap varies the amount of feedback to the base of the transistor with the feedback signal developed between points A and C. As the tap is moved closer to point A the amount of feedback is reduced.

Starting bias is given by R1, R2 and sliding bias is produced by the action of C3, R3. When oscillations have started the oscillatory signal between points A and C, which is in antiphase with the voltage between B and C, is fed back to the base via C2 (d.c. block). The feedback signal receives a phase reversal between base and collector and thus arrives back at point B in phase to sustain the oscillation.

An alternative Hartley oscillator circuit is given in Fig.6.12. L1, C1 form the oscillatory circuit with C2 completing the a.c. path to the common negative line. The tap on L1 is connected to the emitter and point A is connected to the base via the d.c. blocking capacitor C3. Point B at the other end of the inductor is effectively connected to the collector to a.c. via C2 and C4, both

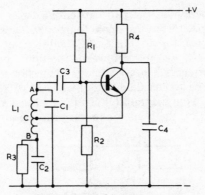

Fig. 6.12 Alternative Hartley oscillator.

of relatively large value. Thus, because of the tap on L1, points A and B are in antiphase. The phase reversal between base and collector of the transistor ensures that the feedback is in phase as is required. Resistors R1 and R2 provide starting bias and C2, R3 sliding bias. R4 is the d.c. feed resistor to the collector.

## Colpitts Oscillator

Another type of L.C. oscillator that is very similar to the Hartley is the Colpitts circuit. Instead of using an inductor with a physical tap, the tap is obtained by using two capacitors. Fig.6.13 shows the basic idea.

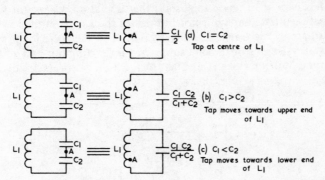

Fig. 6.13 Idea of use of two capacitors to tap coil.

If two capacitors C1 and C2 are connected in series across the inductor L1, the coil will be effectively tapped at a point depending upon the relative values of the capacitors. With equal value capacitors as in Fig.6.13(a) the tap is effectively at the centre of the inductor, i.e. as regards a.c. the potential at the junction of the two capacitors is the same as the tap. Now, the a.c.

potential at the junction of the two capacitors depends upon the relative reactances of the capacitors at the signal frequency. Thus if C1 value is made greater than that of C2, the reactance of C1 will be less than that of C2 and the effective tap on L1 will move towards the upper end of the coil as in diagram (b). Conversely, when C1 value is made less than that of C2, the reactance of C1 will be greater than that of C1 and the tap will move towards the lower end of L1 as in diagram (c). Thus we have a very convenient way of adjusting the tap on the inductor to vary the amount of feedback in an oscillator.

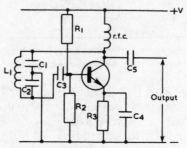

Fig. 6.14 Colpitts Oscillator (shunt fed).

One arrangement of a Colpitts oscillator is shown in Fig.6.14. The oscillatory circuit is formed by L1, C1 and C2, with C1 and C2 tapping the inductor at a suitable point. Because of the tap produced by C1 and C2, oscillatory voltages at either end of L1 will be in antiphase as is required. The transistor provides the usual phase inversion between base and collector so that the feedback is in phase to sustain the oscillation. Capacitor C3 blocks the d.c. between base and collector but passes the a.c. from the lower end of L1 to the base of the transistor. The r.f. choke in the collector prevents the low internal resistance of the d.c. supply from shorting out part of the oscillatory circuit. Starting and sliding bias are as for the previous circuits.

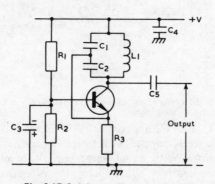

Fig. 6.15 Colpitts oscillator (series fed).

A series fed Colpitts oscillator is given in Fig.6.15. Capacitors C1 and C2 artificially tap the inductor L1 with the tap connected to the emitter of the transistor. As regards a.c. the upper end of L1 is connected to the base via C4 and C3. Thus the base and collector connections at opposite ends of L1 are in antiphase as required. R1, R2 provide the initial starting bias for the transistor. Sliding bias is provided by C3 which charges up with the polarity shown when base current flows on the conducting half-cycles. This causes the operating point to move to class-B or class-C as the oscillation builds up.

**Tuned Load**

When the load to be fed from an oscillator is inductive or tuned, the load can be made to form part or whole of the oscillatory circuit. An example is shown in Fig.6.16.

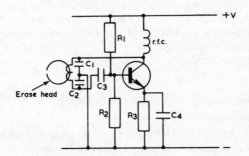

*Fig. 6.16 Load forming part of oscillatory circuit.*

Here the erase head of an audio tape recorder, which is to be fed from the oscillator, is made part of the oscillatory circuit. The inductance L of the erase head together with C1 and C2 are arranged in a Colpitts oscillator circuit. Its operation is identical to that given in Fig.6.14. This idea provides an efficient means of transferring largish amounts of oscillatory power from an oscillator to the load.

**Use of Buffer Stage**

The operating frequency of an oscillator is dependent to some extent upon the magnitude of the load. When the load is reactive, i.e. capacitive or inductive, variations in the loading will affect the frequency of oscillation. The effect of variable loading can be reduced by employing a buffer stage between the oscillator and its load.

A common type of buffer stage is the emitter-follower circuit and an example is given in Fig.6.17. TR1 forms the oscillator transistor connected in

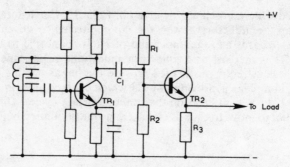

*Fig. 6.17 Use of buffer stage.*

a Colpitts circuit with the output capacitively coupled via C1 to an emitter-follower TR2. Forward bias to class-A is provided by the potential divider R1, R2 and the output is taken to the load from across the emitter resistor R3. An emitter-follower has a high input impedance so there is little loading on the oscillator. Variations in loading occur across R3 in the output circuit of TR2. Because the output impedance of TR2 is low, variations in the loading have little effect on the signal across R3. The use of TR2 thus serves to improve the frequency stability of the oscillator.

## Crystal Oscillators

Some oscillator circuits make use of the piezo-electric property of a crystal in place of an L.C. circuit to determine the frequency of operation. The piezo-electric effect occurs naturally in quartz and can be induced in certain ceramic materials. A slice of quartz cut from a complete crystal in a particular way is given conductive coatings on each side and then sealed in a glass/metal container to protect it from damage and contamination.

The operating frequency of a crystal depends upon the size of the slice; the thinner it is the higher the operating frequency. Crystals are generally available in the range of about 4 kHz to 10 MHz. For higher frequencies, the crystal slice becomes very thin and fragile. The frequency stability of a crystal

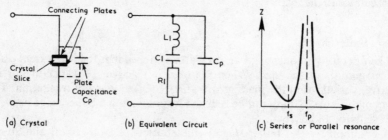

*Fig. 6.18 Piezo electric crystal.*

is of very high order. For example, a 1 MHz crystal may be stable to within 1 Hz over long periods. For high accuracy the crystal should be maintained at a constant temperature, e.g. in a temperature-controlled oven. One important use of crystal oscillators is in transmitters where the carrier frequency must be maintained to a high degree of accuracy. If a carrier frequency above 10 MHz is required, the output from the crystal may be multiplied using a suitable frequency multiplier circuit until the desired frequency is obtained.

A crystal and its equivalent circuit are given in Fig.6.18. From the equivalent circuit it will be seen that a crystal can be maintained in either series or parallel resonance. Series resonance is produced by L1, C1 and parallel resonance by $C_p$ and the inductance of the series arm above its series resonant frequency. In practice the series and parallel resonant frequencies ($f_s$ and $f_p$) are quite close, differing only by about 1 % or less.

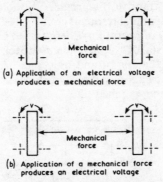

(a) Application of an electrical voltage produces a mechanical force

(b) Application of a mechanical force produces an electrical voltage

·Fig. 6.19 The Piezo electric effect.

The piezo-electric effect is illustrated in Fig.6.19. If a slice of the crystal is subjected to an alternating voltage as in diagram (a), a mechanical force is set up which changes direction in sympathy with the polarity of the applied voltage. Thus the application of an alternating voltage gives rise to a mechanical vibration. The effect is reversible in nature, i.e. if the crystal slice is subjected to a mechanical vibration a voltage is produced between opposite faces of the crystal with a polarity depending upon the direction of the applied force, see diagram (b).

In an oscillator, the crystal is supplied with electrical energy and if the frequency of the energy is close to the natural resonance of the crystal, a very strong mechanical vibration will be set up in the crystal. This vibration will cause a large voltage to be developed between its opposite faces (larger than the applied voltage). Thus the Q of a crystal is very large: about 20,000 as opposed to, say, 200 for an L.C. circuit.

When used in oscillator circuits a crystal may be arranged to operate in its series resonant mode, at a frequency where it presents an inductive reactance or in its parallel resonant mode. An example of the series resonant mode is given in Fig.6.20.

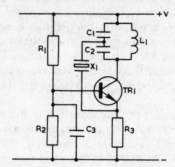

Fig. 6.20 Crystal oscillator using series resonant mode.

Starting bias for the transistor is provided by R1, R2. The collector load for the transistor is the parallel resonant circuit L1, C1 and C2 which is tuned to the desired frequency of operation. Energy is fed back from the collector to the emitter via the crystal X1. Maximum feedback will occur at the series resonant frequency of the crystal and only then will the loop gain be sufficient to sustain the oscillation. Because of the high Q of the crystal a very stable oscillation frequency is achieved.

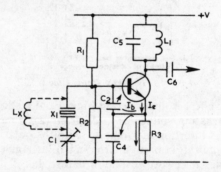

Fig. 6.21 Oscillator with crystal presenting an inductive reactance.

Another crystal oscillator circuit is given in Fig.6.21. In this example the crystal is operating just above series resonance so that it presents an inductive reactance. The crystal may therefore be replaced by an inductance $L_x$ (shown dotted). The circuit is then more readily recognizable as a Colpitts oscillator with C2 and C4 tapping the inductor $L_x$. By adjusting C1, the frequency of oscillation may be altered slightly (a vari-cap diode may be included in series with the crystal to give fine adjustment over the frequency).

A portion of the oscillatory signal appearing between the base and the negative line is developed across C2 which provides the input current Ib. Due to transistor action Ib is amplified and feedback to the oscillatory circuit is provided by the larger portion of Ie that flows in C4. Since Ib and Ie are in phase, the feedback is positive and thus will maintain the oscillation. Resistors R1 and R2 provide starting bias. L1, C5 in the collector circuit is used to reject harmonics at the output.

### R–C Sinewave Oscillators

When a low frequency sinewave in the range of about 10 Hz to 100 kHz is required, a Resistance–Capacitance (R–C) oscillator is normally used. This is because LC oscillators are difficult to design at low frequencies as L and C values become large. There are two possible arrangements for an RC oscillator and the basic principle of both types is illustrated in Fig.6.22.

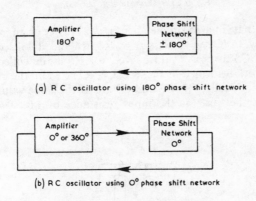

(a) R C oscillator using 180° phase shift network

(b) R C oscillator using 0° phase shift network

*Fig. 6.22 Basic principle of R.C. oscillators.*

In diagram (a) the output of an amplifier feeds an R.C. phase shift network which introduces a phase shift of $\pm180°$. The output from this circuit is then fed back to the input of the amplifier. If the amplifier (usually a single stage) introduces phase inversion of its input signal, the output of the amplifier will arrive back to the network in phase, i.e. the overall phase shift is 0° or 360°. If the oscillator is to give a sinewave output the phase shift introduced by the phase shift network must change with fréquency. However, since the output from the phase shift network will only be in antiphase with its input at one frequency, oscillation will occur at this single frequency. For oscillations to take place, the amplifier must make up for the losses introduced by the phase shift network, i.e. the loop gain must be equal to or greater than unity.

With the other arrangement shown in diagram (b) an amplifier is used which does not provide phase inversion of its input signal, i.e. a two stage amplifier. Since the phase shift network used here does not produce any phase shift at the desired operating frequency, the overall phase shift is 0° or 360° as is required for oscillations to be established. Again, the amplifier must make up for the losses incurred in the network.

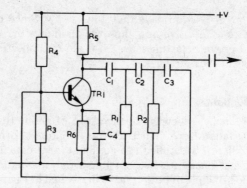

*Fig. 6.23 R-C phase shift oscillator.*

An oscillator that is based on the idea Fig.6.22 (a), is given in Fig.6.23. The transistor is forward biased to class-A by R3, R4 and stabilised by R6 which is decoupled by C4 to prevent n.f.b. Resistor R5 is the collector load. The phase shift network is formed by C1, R1; C2, R2; C3 and the input resistance of the transistor. Usually, C1, C2 and C3 are of the same value with R1 and R2 values made the same as the input resistance of TR1 ($R_{IN}$). Each CR

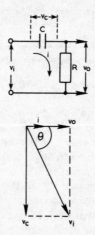

*Fig. 6.24 Phase shift ($\theta$) introduced by isolated RC section.*

section introduces a leading phase shift as illustrated in Fig.6.24 but the three C.R. sections must provide an overall phase shift of 180° at a single frequency. In practice, the calculation of phase shift is complicated by the fact that each R.C. section tends to load the preceding one. If variable frequency operation is required, C1, C2 and C3 may be made variable and ganged together. The approximate frequency of oscillation is given by

$$f_o = \frac{1}{2\pi CR\sqrt{6}} \quad \text{Where } C = C1 = C2 = C3 \text{ and } R = R1 = R2 = R_{IN} = R5$$

$$= 0 \cdot 065 \frac{1}{CR}$$

At $f_o$ the loss in the phase shift network is 29 and thus the $h_{fe}$ of the transistor must be greater than 29. Since the frequency is proportional to $\frac{1}{C}$, a given capacitance change will produce a larger frequency variation than with an LC oscillator where frequency is proportional to $\frac{1}{\sqrt{C}}$.

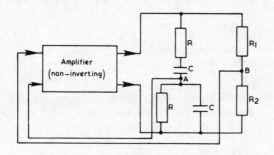

*Fig. 6.25 Wein bridge R-C oscillator.*

A circuit using the principle of Fig.6.22 (b) is given in Fig.6.25. This circuit makes use of a Wein bridge which is shown external to the amplifier. The output from the Wein bridge is taken between points A and B and applied back to the input of a non-inverting amplifier, i.e. a two-stage common-emitter amplifier. When the bridge is balanced the output will be zero and the frequency at balance is given by

$$f_o = \frac{1}{2\pi CR}.$$

In order to produce oscillations, the bridge is operated slightly off balance by varying the ratio of R1 to R2:

$$\text{at balance } \frac{R_2}{R_1 + R_2} = \frac{1}{3}$$

When operated slightly off balance, the output between points A and B is in phase with the input to the network and the overall feedback is positive as is required. Only at one frequency will the phase lead, introduced by the series R.C. equal the phase lag produced by the parallel R.C. to give zero phase shift. For best operation the output resistance of the amplifier should be low and its input resistance high. The voltage gain required by the amplifier to maintain oscillation should be greater than 3.

To produce a good sinewave output from an RC oscillator, the loop gain should be made nearly to unity. If it is greater, the transistor will be driven into the non-linear part of its characteristic and the output will be distorted. The output waveform can be improved by incorporating a control in the amplifier to vary the gain. Alternatively, automatic amplitude stabilisation may be employed. This may be achieved in the Wein bridge oscillator by replacing R1 with a n.t.c. thermistor. Thus, as the amplitude of the oscillation increases, the voltage across the thermistor increases which causes its resistance to fall and for the bridge to come closer into balance. As a result the output from the bridge is reduced and the amplitude of oscillation is kept constant.

## QUESTIONS ON CHAPTER SIX

(1) An LC or RC oscillator circuit will oscillate if:
(a) Loop gain is equal to unity and overall feedback is in antiphase
(b) Loop gain is less than unity and overall feedback is in phase
(c) Loop gain is less than unity and overall feedback is in antiphase
(d) Loop gain is equal to unity and overall feedback is in phase.

(2) Resistance present in an L.C. oscillatory circuit will result in:
(a) An increase in the circuit Q
(b) Loss of energy
(c) Less feedback being required
(d) Lower harmonic level in the output.

(3) Sliding bias is used in an L.C. oscillator to:
(a) Keep the oscillator amplitude constant
(b) Make the oscillator self-starting
(c) Increase the forward bias as the oscillation builds up
(d) Bias the transistor to class-A.

(4) To increase the frequency of an L.C. oscillator:
(a) C may be increased
(b) L may be increased
(c) C may be decreased
(d) L is increased and C decreased by the same percentage.

Questions 5–7 refer to Fig.6.6 on page 116

(5) If R1 goes open circuit the effect will probably be:
(a) Smaller amplitude of oscillation
(b) Larger amplitude of oscillation
(c) Wrong frequency of oscillation
(d) Oscillation will fail to start.

(6) If the connections to L2 are reversed the effect will be:
(a) Oscillation will fail to start
(b) Nil
(c) Larger amplitude of oscillation
(d) Change in oscillator frequency.

(7) If L1 goes open circuit the effect will be:
(a) Oscillation at low frequencies only
(b) No oscillation
(c) Oscillation at high frequencies only
(d) Intermittent operation.

(8) A circuit that would be suitable for use as a local oscillator in an a.m. radio would be:
(a) A crystal oscillator
(b) A Reinartz oscillator
(c) A series fed Hartley oscillator
(d) A Wein bridge oscillator.

(9) An oscillator that would be suitable for generating a 15 Hz sinewave would be:
(a) A crystal oscillator
(b) A Hartley oscillator
(c) A Colpitts oscillator
(d) An R.C. oscillator.

(10) The most suitable oscillator for generating the carrier of a broadcast radio transmission would be:
(a) A crystal type
(b) An R.C. type
(c) A Hartley type
(d) A Colpitts type.

(11) The main use of a buffer stage with an oscillator is to:
(a) Increase the amplitude of the oscillator output
(b) Improve the frequency stability
(c) Provide a high impedance source to the load
(d) Allow ease of tuning.

(12) Series capacitors are used to tap the oscillatory circuit inductor in:
(a) A Hartley oscillator
(b) A Reinartz oscillator
(c) A Colpitts oscillator
(d) A tuned base oscillator.

# CHAPTER SEVEN

# MULTIVIBRATORS

THERE ARE OSCILLATORS that are specifically used to generate non-sinusoidal waveforms. Multivibrators are in this class of oscillator, and are employed to generate rectangular waveforms. In this chapter three types of multivibrator circuit will be considered.

(a) Astable Multivibrator
(b) Monostable Multivibrator (also known as the Flip-Flop)
(c) Bistable Multivibrator.

**Transistor as a Switch**

Multivibrator oscillators use two transistors which operate as electronic switches. The use of a bipolar transistor as a switch is illustrated in Fig.7.1. A transistor may be switched rapidly on and off by the application of a rectangular pulse to its base as shown in diagram (a). When the input voltage is below the level required of, say, 0·7V, to produce collector current flow, the transistor is off and only a small leakage current (Iceo) flows in the collector load $R_L$. Thus the voltage drop across $R_L$ is negligible and the Vce of the transistor is almost equal to the supply line voltage $V_L$. When the base voltage rises above 0·7V, the transistor conducts and if supplied with sufficient base current the collector current saturates. Under this condition the collector voltage falls to a small value and is said to bottom.

The fully off and fully on conditions are best seen from the load line given in diagram (b). The load line is drawn from a point on the Vce axis corresponding to the supply voltage and with a slope appropriate to the value of $R_L$. It will be seen that if the base current is equal to or greater than 40 $\mu$A, the operating point is at A. This is the fully on condition where the collector current saturates and the voltage drop across the transistor falls to a low value or bottoms. The bottoming voltage is only about 0·25V, depending upon the type of transistor used. In the fully off condition the operating point is point B where the base current is zero. An f.e.t. will operate as a switch almost identically but with a slightly higher bottoming voltage of 0·3 to 1·0V and a smaller leakage current of 1 nA.

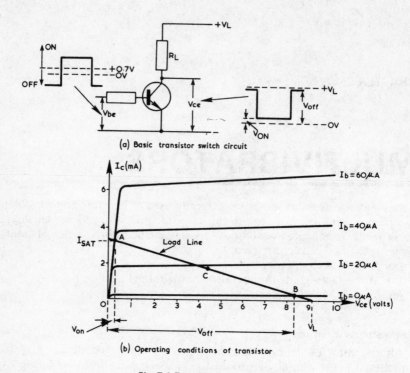

(a) Basic transistor switch circuit

(b) Operating conditions of transistor

Fig. 7.1 Transistor as a switch.

The power dissipated within the transistor is small at points A and B. When the transistor is fully on there is only a small voltage drop across it (although the current is relatively large) and when fully off only a small leakage current flows (although the volts drop across the transistor is equal to the supply line voltage). Between the two extreme points on the load line, the transistor operates at various points as it changes over from the off to on condition. At point C, halfway up the load line, the dissipation is greater than at the extremes and is at a maximum. Thus if the transistor is to act as a switch, the switching time between A and B should be short so that the dissipation is small. As the power dissipation is small in the fully on and fully off conditions quite small power transistors can be used for high-speed switching.

It would be expected that at the instant the base current is reduced to zero when switching from on to off the collector current would immediately stop flowing. In practice, it continues to flow for a short time due to what is known as charge storage. This occurs in the base region of the transistor and acts like a charged capacitor. Charge storage causes a delay in the cut-off time of the transistor, depending on its type and construction.

## Astable Multivibrator

An astable multivibrator is a free-running oscillator consisting of two cascaded RC coupled stages connected to give 100% positive feedback. A basic circuit is given in Fig.7.2 where the outputs from the collectors of TR1 and TR2 are cross-coupled to the base of the other transistor via the coupling capacitors C1 and C2. Resistor R1 provides the base bias for TR2, and R2 provides the base bias for TR1. Resistors R3 and R4 are collector load resistors.

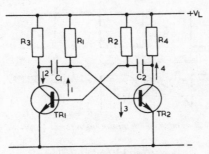

Fig. 7.2 Basic astable multivibrator.

Assuming that R3=R4, R1=R2, C1=C2, and similar transistors are used to produce a symmetrical circuit, it may be thought that both transistors would conduct together and remain conducting since they are both biased on via their base resistors R1 and R2. In practice, however, immediately following the connection of the supply voltage, one transitor will become fully conducting and the other fully off. This is due to the cross-coupling and it is quite random which transistor becomes fully conducting initially.

Any disturbance such as a noise voltage or inequality in the two transistors (their d.c. current gains will not be identical) will produce the initial condition of one transistor conducting hard and the other being cut-off. For example, consider a noise voltage disturbance at TR1 base acting in the positive direction and indicated by arrow 1. Now, a rise in voltage at TR1 base will be amplified by TR1 to produce a fall in voltage at its collector (arrow 2). This fall in voltage is passed via C1 to TR2 base (arrow 3), reducing the current in TR2 and causing its collector voltage to rise (arrow 4). The rise in voltage is then passed via C2 back to TR1 base (note that it arrives back in phase with the original rise, i.e. positive feedback). This rise will receive repetitive amplification in TR1 and TR2 until TR2 is cut off by the large fall in voltage at its base and TR1 is conducting hard due to the rise in voltage at its base. This accumulative action takes place very quickly at switch on. However, the circuit does not remain in this state for very long which is explained with the aid of the waveforms in Fig.7.3.

At time $t_o$, TR1 is fully on with its collector voltage bottomed, and TR2 is fully off with its collector voltage at practically the full supply line potential

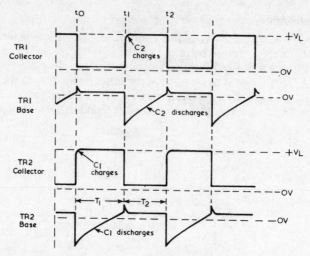

Fig. 7.3 Astable multivibrator waveforms ($C_1 R_1 = C_2 R_2$).

$V_L$. TR2 is held off by the negative voltage at its base.

The following reactions then occur:

(1) Since the voltage at the lower end of R1 (w.r.t. the negative line) is negative by a voltage almost equal to $V_L$, and the upper end of R1 is connected to $+V_L$, there is a voltage across R1 of twice the supply line potential. This voltage causes a current to flow in R1 thereby allowing C1 to discharge. As C1 discharges exponentially via R1, the base voltage of TR2 rises. When it has risen just a little above zero, TR2 starts to conduct.

(2) The conduction of TR2 causes its collector voltage to fall and this fall is passed via C2 to TR1 base. As a result, the conduction in TR1 is reduced and its collector voltage rises. The rise in voltage is applied via C1 to TR2 base, increasing the conduction of TR2. As a result, TR2 collector voltage falls even further . . . and so on. The accumulative action due to positive feedback very quickly results in TR2 coming hard on and TR1 switching off. TR1 is held off due to the negative voltage at its base. Thus at instant $t_1$, TR2 collector voltage bottoms and TR1 collector voltage rises to $+V_L$.

(3) There is now a voltage established across R2 that is equal to twice the supply line potential. This voltage causes a current to flow in R2 thereby allowing C2 to discharge. As C2 discharges exponentially via R2, the base voltage of TR1 rises. When it rises a little above zero, TR1 starts to conduct.

(4) The conduction of TR1 causes its collector voltage to fall and this fall is passed via C1 to the base of TR2. As a result, the conduction of TR2 is reduced and its collector voltage rises. The rise in voltage is applied via C2 to TR1 base increasing the conduction of TR1 even more. Again, due

to the positive feedback loop, TR1 becomes fully conducting very rapidly and TR2 quickly switches off. TR1 collector voltage is then bottomed and TR2 collector voltage is equal to the line supply potential $+V_L$.

This completes a full cycle of events at instant $t_2$ and the cycle will repeat itself continuously. It will be seen that neither TR1 nor TR2 remain in a stable state for long periods, hence the name astable (not stable). Rectangular voltage waveforms are available as outputs from either collector, with TR1 collector being in antiphase with TR2 collector.

The period $T_1$ that TR2 is off is determined by the time-constant of C1.R1 and the off period for TR1 ($T_2$) is determined by the time constant C2.R2. The periodic time T of the waveform is given by:

$$T = 0.69 \ (C1.R1 + C2.R2) \quad s$$

or the frequency f by

$$f = \frac{1}{0.69(C1.R1 + C2.R2)} \ Hz$$

## Asymmetrical Output

When the time constants C1 R1 and C2 R2 are equal, a symmetrical or square-wave output is obtained as shown in Fig.7.3. If the time constants are unequal an asymmetrical wave (unequal mark-to-space ratio) will be produced.

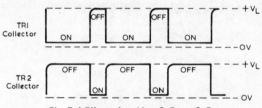

Fig. 7.4 Effect of making $C_1R_1 > C_2R_2$.

An example is shown in Fig.7.4 where C1R1 > C2R2. Under this condition, TR1 will be fully on for relatively long periods and off for relatively short periods. The converse is true for TR2. An asymmetrical output may be required for instance when it is to be used to produce a sawtooth, as in a timebase.

## Variable Frequency Operation

The frequency of an astable multivibrator may be altered by adjusting the time constants of C1 R1 and C2 R2. Shortening either time constant will increase the frequency, and lengthening the time constant of either will lower the frequency. However, if only one time constant is altered the mark-to-space ratio will be changed which may be undesirable. To maintain a constant mark-to-space ratio as the frequency is altered, the arrangement shown in Fig.7.5 may be used.

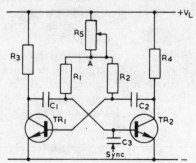

Fig. 7.5 Variable frequency operation.

The base bias resistors R1 and R2 are returned via a common resistor R5 to the supply line. One may consider the effect of varying R5 as altering the aiming potential for both time constants, which is illustrated in Fig.7.6.

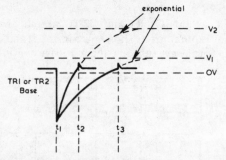

Fig. 7.6 Effect of varying the aiming potential with $R_5$.

Since both time constants are equally affected, only one base waveform need be considered. Suppose that the aiming potential at point A as set by R5 is $V_1$ volts. If a coupling capacitor commences its discharge at instant $t_1$ the base voltage will rise exponentially aiming for $V_1$. However, as soon as the voltage rises just above zero at instant $t_3$ the precipitate action of one transistor coming on and the other going off commences. Suppose now that the resistance of R5 is reduced to increase the aiming potential at point A to $V_2$ volts. When a coupling capacitor commences its discharge at instant $t_1$ it will aim for $V_2$, thus the initial slope of the discharge curve is steeper. In consequence, the base voltage will reach the cut on point for the transistor in a shorter period of time, i.e. at instant $t_2$. The shorter the change over period, the higher will be the frequency of operation. Thus the higher the aiming potential, the higher is the frequency of p.r.f. The variable resistor R5 may therefore serve as a frequency control. In some circuits R5 may be returned to a higher potential than the line supply voltage; this provides steep discharge curves and better stability.

An interesting situation arises if the aiming potential is a sinewave. This would cause frequency modulation of the multivibrator; a technique that is used in some video cassette recorders.

## Synchronisation

The frequency stability of an astable multivibrator is not high, but it can easily be locked or synchronised to an external source of the same frequency or a frequency which is a multiple of its free-running frequency.

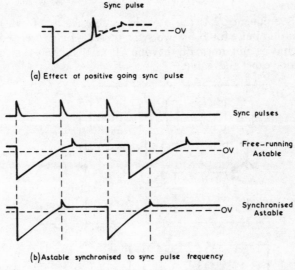

Sync pulse

(a) Effect of positive going sync pulse

Sync pulses

Free-running
Astable

Synchronised
Astable

(b) Astable synchronised to sync pulse frequency

Fig. 7.7 Synchronisation of astable.

Synchronisation may be achieved by feeding positive going pulses to the base of one of the transistors as in Fig.7.5 where the sync. pulses are supplied to TR2 base via C3. The effect of a sync. pulse on the base voltage waveform is shown in Fig.7.7(a). If the sync. pulse arrives at the base when that transistor is on it will have no effect since the transistor will be bottomed. However, if it arrives when the transistor is off it will cause the base voltage to rise above zero in advance of the instant where, unaided, the transistor would normally reach the cut-on state as shown. To bring the astable exactly into synchronism with the external sync. pulses, the free-running frequency must be adjusted so that it is slightly lower than the frequency of the sync. pulses as shown in diagram (b).

Should it be desirable to lock the astable circuit to every other sync. pulse as in Fig.7.8, the free-running frequency should be made slightly lower than

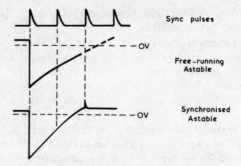

*Fig. 7.8 Astable synchronised to sync pulse frequency ÷ 2.*

the sync. pulse frequency/2. In this way, the astable acts as a frequency divider, giving one output pulse for every two sync. pulses. Higher frequency division is possible, but is not normally beyond 10 as it is then difficult to make synchronisation exact and stable.

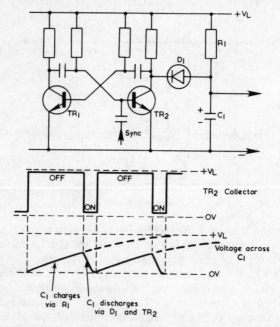

*Fig. 7.9 Using an astable and CR network to produce a sawtooth.*

## Use of Astable to Produce a Sawtooth

A simple way of producing a free-running timebase is to use a CR sawtooth forming network and an astable multivibrator as shown in Fig.7.9. Suppose that the mark-to-space ratio of the astable is set so that TR2 is off for long periods and on for relatively short periods as shown by the waveform for TR2 collector.

When the transistor is off, D1 is off and C1 charges with the polarity shown towards $+V_L$. If the time constant C1,R1 is made long compared with the TR2 off period, the rise in voltage across C1 will be fairly linear. As soon as TR2 conducts, and its collector voltage bottoms, D1 will become forward biased allowing C1 to rapidly discharge via D1 and TR2. If the on period of TR2 is long compared with the discharge time of C1, the voltage across C1 will fall to zero. Thus, when TR2 goes off again, C1 will commence to charge again via R1 thereby producing a repetitive sawtooth voltage waveform across C1.

The frequency of the sawtooth waveform is determined by the frequency of the astable multivibrator and this may be locked to an external sync. source if desired. The rise in voltage across C2 corresponds to the scan of the timebase waveform and the fall in voltage to the flyback section. The longer the time-constant of C1,R1 in relation to the off time of TR2, the more linear will be the scan but the smaller its amplitude.

## Monostable Multivibrator

A monostable oscillator is a triggered oscillator, i.e. not free-running and only has one stable state. At switch-on the circuit takes up this state with one transistor fully on and the other fully off. Upon the application of a trigger pulse the circuit switches over and the conducting states of the transistors are reversed. However, this is not a stable state and after a period of time determined by the circuit components it reverts to its former stable state. It remains in this state until another trigger pulse is applied.

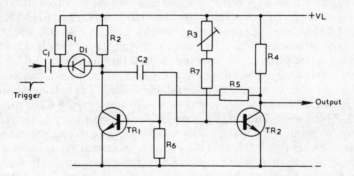

*Fig. 7.10 Monostable multivibrator and trigger circuit.*

A typical collector coupled circuit is given in Fig.7.10. It will be seen that there is a.c. coupling between TR1 and TR2 via C2, but d.c. coupling between TR2 and TR1 via R5. At switch on, TR2 conducts hard as its base is supplied from the positive supply line via R3 and R7. Thus TR2 collector voltage is bottomed and low at, say, 0·25V. By suitable choice of R5 and R6 values, the voltage at TR1 base will be insufficient to bias TR1 on. Therefore, at switch on· TR1 is off and TR2 is on; this is the stable state. It is assumed here that silicon transistors are used, but in circuits using germanium types the lower end of R6 may be taken to a negative potential to ensure that TR1 is fully off. As TR1 is off its collector potential will be high and C2 will be charged to practically the full line supply potential. These are the conditions prevailing prior to instant $t_1$ in the waveforms of Fig.7.11. Trigger pulses are fed to TR2 base via C1 and the trigger diode D1. Since TR1 is off, both sides of the diode are at the positive supply voltage. Thus, on receipt of negative trigger pulses at D1 cathode, the diode conducts and a negative going pulse is produced across R2 which is fed to TR2 base via C2.

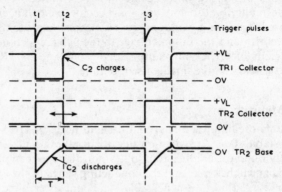

Fig. 7.11 Waveforms of monostable multivibrator.

The negative pulses applied to TR2 base at instant $t_1$ reduce its collector current, causing its collector voltage to rise. This rise is reflected at TR1 base causing TR1 to commence conducting. In consequence, TR1 collector voltage falls and this is passed to TR2 base via C2 turning TR2 further off. This action is repeated quickly and often resulting in TR2 turning sharply off and TR1 coming quickly to the fully on state. TR2 is held off by the negative potential at TR2 base (in the same way as in the astable circuit). The voltage established across R3 and R7 now causes a current to flow discharging C2 through R7 and R3. This causes TR2 base voltage to rise exponentially towards $+V_L$. However, as soon as the voltage rises a little above zero TR2 commences to conduct. The resultant fall in TR2 collector potential commences an accumulative action that results in TR1 rapidly switching off and TR2 quickly coming hard on. The circuit is now back in its original stable state and will remain so until the next trigger pulse is applied. Note that when TR1 is conducting, D1 is reversed biased, thus the fall in TR1 collector potential is not fed back into the trigger pulse source.

It will be seen that at the collectors of the transistors, pulse waveforms are available having a duration T. This interval is essentially governed by the time-constant C2, R7 and R3. By varying the time-constant with the aid of R3, the duration of the pulse can be varied over quite wide limits when suitable component values are employed. Note that the p.r.f. of the monostable is determined by the p.r.f. of the trigger pulses.

### Emitter Coupled Monostable

There are a number of advantages in replacing the direct coupling between TR2 collector and TR1 base with emitter coupling. A basic circuit is given in Fig.7.12 where the direct coupling has been replaced by emitter coupling via the common emitter resistor R8.

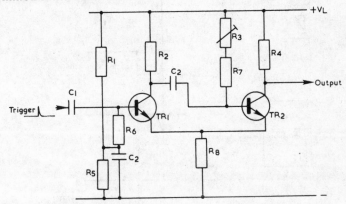

*Fig. 7.12 Emitter coupled monostable multivibrator.*

In the stable state, TR2 is conducting as its base is returned to the positive supply line via R7,R3. The current due to TR2 flowing in R8 makes the emitter of TR1 positive and by suitable choice of values for R1 and R5, TR1 can be arranged to be cut off. When a positive trigger pulse is fed through C1 to TR1 base, the transistor conducts and its collector voltage falls. The fall in voltage is passed through C2 to TR2 base causing TR2 to turn off. As there is now no emitter current in TR2, the voltage across R8 falls a little which maintains TR1 in the on state. Capacitor C2 now discharges through R7 and R3 and when TR2 base voltage rises a little above the voltage across R8, TR2 conducts once more. As a result of this, the voltage across R8 rises which turns TR1 off. The circuit is now back in its original stable state.

As with the collector coupled circuit, the pulse duration at the output is settled by the time constant C2, R7 and R3 and the p.r.f. of the monostable by the p.r.f. of the trigger pulses. This arrangement has the following advantage:
(a) The output may be taken from TR2 collector which is now isolated from the coupling path between transistors.
(b) Similarly, the trigger circuit is isolated from the feedback path as the trigger is applied direct to TR1 base.

(c) Greater flexibility is allowed in choosing the operating points of the transistors which is of importance in high-speed switching applications.

## Use of Monostable

The main use of a monostable multivibrator is to produce a pulse of desired duration and/or a pulse with a fixed time delay. An example is given in Fig.7.13.

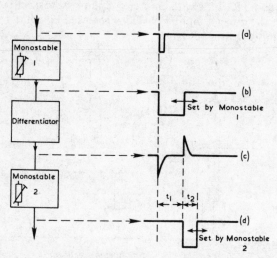

*Fig. 7.13 Use of Monostables to produce a delayed pulse.*

Suppose that is required to produce a pulse with a particular duration having a fixed time delay from the leading edge of the pulse shown at (a). Now, if the pulse at (a) is used to trigger Monostable 1, the pulse output shown at (b) will have a width dependent upon the time-constant of this monostable and this can be set to give the duration $t_1$. If the output of Monostable 1 is then fed to a differentiating circuit, the output will be as at (c). By using the positive-going spike of waveform (c) to trigger Monostable 2, its output will be as at (d) with a duration $t_2$ set by the time constant of the monostable.

Thus, at (d) we have a pulse with a fixed time delay $t_1$ from the commencement of the pulse at (a) and with a particular duration $t_2$.

## Bistable Oscillator

This type of multivibrator is extensively used in counter circuits and its basic principle was described in *Electronic Systems*. It is now necessary to study its circuit action, a typical circuit being given in Fig.7.14.

As the name bistable implies this multivibrator has two stable states with one transistor conducting and the other cut off, producing one of its stable states. Upon the receipt of a trigger pulse the circuit changes over; the transistor that was conducting is now cut off and the transistor that was off is now conducting. This is the second stable state. The next trigger pulse to be applied causes the bistable to revert to its original stable state. A bistable is thus a triggered multivibrator.

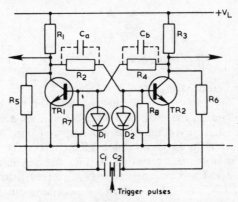

*Fig. 7.14 Bistable oscillator.*

The two transistors TR1 and TR2 are cross-coupled by R2 and R4 so that positive feedback occurs. Both transistors cannot be conducting simultaneously, except for a brief period during changeover. Thus, initially when the supply is connected one transistor is conducting and the other is cut off.

Suppose that TR1 is conducting and TR2 is cut off. As a result TR1 collector will be bottomed and, due to the low collector voltage of, say, 0·25V, the base voltage of TR2 will be insufficient to turn TR2 on (silicon transistors are assumed). Thus TR2 collector voltage will be at $+V_L$. This is the condition illustrated in the waveforms of Fig.7.15 just prior to instant $t_1$.

To change over the circuit conditions a negative going trigger pulse is required to be fed to TR1 base to switch TR1 off. The trigger pulses are applied through C1 and C2. To ensure that each trigger pulse is applied to the appropriate transistor, steering diodes D1 and D2 are employed. When TR1 is on, D1 is biased on but D2 is off. The first trigger pulse to arrive is steered through D1 to TR1 base. Here it causes TR1 to start to turn off. The resulting rise of voltage at TR1 collector is passed via R2 to TR2 base causing TR2 to conduct. As a result TR2 collector voltage starts to fall which is passed via R4 to TR1 base causing TR1 current to reduce even further. This action is repeated rapidly many times resulting in TR1 quickly going off and TR2 rapidly coming hard on (instant $t_1$).

Thus the state of the circuit has been changed with TR1 cut off and TR2 conducting. The circuit will remain in this state until the next trigger pulse

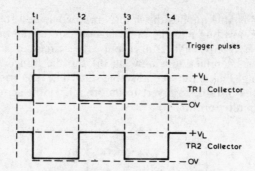

Fig. 7.15 Bistable oscillator waveforms.

arrives. This pulse will be steered through D2 which is now biased on to TR2 base (D1 is off). TR2 now starts to turn off and the rise in voltage at its collector is passed on to the base of TR1 via R4. This action causes TR1 to turn on and the fall in voltage at its collector is passed to TR2 base via R2 where the effect is to further reduce TR2 current. This action is also repeated rapidly many times resulting in TR1 coming hard on and TR2 going off (instant $t_2$). The next trigger pulse is steered through D1 causing TR1 to cut off and TR2 to conduct (instant $t_3$) . . . and so on.

It will be noted that for every two trigger pulses, one complete cycle of bistable action takes place, i.e. the p.r.f. of the output at either collector is half the p.r.f. of the input trigger. Thus a bistable may be used as a $\div 2$ stage.

### Speed-up Capacitors

In some bistable oscillator circuits, R2 and R4 are shunted with small value capacitors (Ca and Cb). These are called speed-up capacitors, and the need for them may be seen from Fig.7.16.

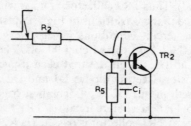

Fig. 7.16 Need for speed-up capacitor.

When TR2 is off it may be regarded that R5 is shunted by $C_i$ (the capacitance of the reverse biased base-emitter junction of TR2). When TR1 goes off and its collector rises as shown, R2 and $C_i$ form a low pass filter for the step-voltage change. As a result the base waveform of TR2 would be rounded as indicated (showing the loss of high frequency components) and this would increase the switching time of the circuit. If R2 is shunted with a capacitor of suitable value, the resistor will be by-passed at h.f. and the switching time of the circuit will be reduced. Since the same argument holds good when TR1 is off, it is also necessary to shunt R4 with a speed-up capacitor.

## QUESTIONS ON CHAPTER SEVEN

(1) When a transistor is used as a switch and is in the fully on condition, the collector-to-emitter voltage with a 9V supply line will probably be:
(a) 9V
(b) 4·5V
(c) 2·2V
(d) 0·25V.

(2) Which of the following would result in a decrease in the frequency of the astable multivibrator circuit of Fig.7.2 on page 135:
(a) Smaller C1 value
(b) Larger C2 value
(c) Smaller C2 value
(d) Larger R4 value.

(3) Which of the following components faults would result in TR1 collector potential in Fig.7.2 being permanently at $+V_L$ volts:
(a) R2 o/c
(b) R1 o/c
(c) R4 o/c
(d) R3 o/c.

(4) In Fig.7.2 if the capacitor C1 goes open circuit the effect will be:
(a) P.R.F. will increase
(b) Asymmetrical output
(c) No output
(d) Sawtooth output.

(5) In Fig.7.2 if the capacitor C2 goes open circuit, the collector-to-emitter voltages of TR1 and TR2 with a supply line of 10V will be:
(a) TR1 0·25V and TR2 10V
(b) TR1 10V and TR2 0·25V
(c) TR1 0·25V and TR2 0·25V
(d) TR1 10V and TR2 10V.

(6) A d.c. voltmeter connected between the emitter and collector of TR1 or TR2 in Fig.7.2, when the circuit is producing a square-wave output from a 10V supply, will read about:
(a) 0·25V
(b) 0·5V
(c) 5V
(d) 9·25V.

(7) There is no output from the circuit in Fig.7.10 and the collector-to-emitter voltage of TR2 is 0·3V with a supply rail voltage of 10V. Which of the following component faults could cause these symptoms:
(a) R3 o/c
(b) TR2 base-emitter short
(c) C1 o/c
(d) TR2 emitter connection o/c.

(8) Which of the following is a free-running oscillator:
   (a) An astable multivibrator
   (b) A bistable multivibrator
   (c) A monostable multivibrator
   (d) A flip flop.

(9) The p.r.f. of the trigger applied to a bistable is 10kHz. The output waveform from one of the transistors will be at a p.r.f. of:
   (a) 10 kHz
   (b) 20 kHz
   (c)  1 kHz
   (d)  5 kHz.

(10) The p.r.f. of the trigger pulses applied to a monostable is 500Hz. The output pulse will have a p.r.f. of:
   (a) 250 Hz
   (b) 500 Hz
   (c)   1 kHz
   (d)  50 Hz.

CHAPTER EIGHT

# LOGIC CIRCUITS

IN VOLUME 1 of this series basic logic gates were introduced and examples of simple gate combinations were given. It is now necessary to consider gate combinations further and to develop more complex logic circuits.

**Positive and Negative Logic**

In digital circuits the logic 1 and 0 states correspond to particular electrical voltages. If the **more positive** of the two states is selected as logic 1 then the system is said to use **positive logic**, see Fig.8.1(a). On the other hand if the **more negative** of the two states is selected as logic 1, the system is said to use **negative logic**, see Fig.8.1(b).

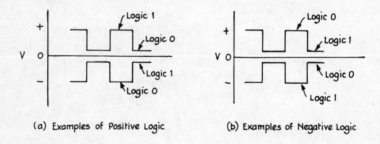

(a) Examples of Positive Logic          (b) Examples of Negative Logic

*Fig. 8.1 Positive and negative logic.*

With logic families, e.g. TTL, CMOS, I²L, etc. the actual logic voltage levels found in digital circuits fall within stated voltage ranges. There must be sufficient voltage clearance between these ranges so that there is no ambiguity in recognising a logic 1 voltage from a logic 0 voltage; see Fig.8.2 which illustrates the voltage ranges for TTL. Here positive logic is assumed since the more positive of the two voltage ranges is designated logic 1. There is no reason why negative logic notation should not be used; however, it should be noted that the

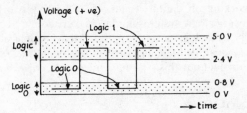

*Fig. 8.2 TTL logic levels (positive logic).*

name given to a logic gate as defined by its truth table depends upon the logic notation employed. Consider the following example:

Suppose that the voltages measured at the input and output of a 2-input logic gate are as shown in the truth table of Fig.8.3. If **positive logic** is assumed where logic 1 = + 5V and logic 0 = + 0.2V, then the truth table may be redrawn as in Fig.8.4. It will be seen that the gate behaves as an AND gate. On the other hand, if **negative logic** is assumed where logic 1 = 0.2V and logic 0 = 5V, the truth table may be redrawn as in Fig.8.5. In this case it will be noted that the gate behaves as an OR gate!

| Inputs | | Output |
|---|---|---|
| A | B | F |
| + 0·2V | + 0·2V | + 0·2V |
| + 0·2V | + 5·0V | + 0·2V |
| + 5·0V | + 0·2V | + 0·2V |
| + 5·0V | + 5·0V | + 5·0V |

*Fig. 8.3 Input & output voltages of 2-input gate.*

| A | B | F |
|---|---|---|
| 0 | 0 | 0 |
| 0 | 1 | 0 |
| 1 | 0 | 0 |
| 1 | 1 | 1 |

*Fig. 8.4 Positive logic (AND function).*

| A | B | F |
|---|---|---|
| 1 | 1 | 1 |
| 1 | 0 | 1 |
| 0 | 1 | 1 |
| 0 | 0 | 0 |

*Fig. 8.5 Negative logic (OR function).*

Thus, when working on digital circuits it is important to be aware of the notation used. Positive logic systems tend to be very common but do not take it for granted that positive logic always applies. **Mixed logic** is also possible, e.g. the inputs may use positive logic and the outputs negative logic. Positive logic will be assumed throughout this chapter, unless otherwise stated.

### EX-OR and EX-NOR Gates

*Exclusive-OR*

An exclusive-OR gate has two inputs and a single output. Its symbol is given in Fig.8.6(a). The output from an exclusive-OR gate will assume the logic 1 state if one and only one input is as logic 1. A truth table for the gate is given in Fig.8.6(b). It will be seen that unlike the OR gate, output is not obtained when both inputs are at logic 1, i.e. it excludes this condition. Since an output is obtained only when the input logic levels are **different**, it is sometimes referred to as a **non-equivalence** gate.

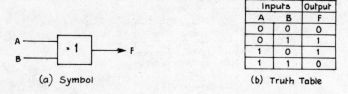

| Inputs | | Output |
|---|---|---|
| A | B | F |
| 0 | 0 | 0 |
| 0 | 1 | 1 |
| 1 | 0 | 1 |
| 1 | 1 | 0 |

(a) Symbol  (b) Truth Table

*Fig. 8.6 Exclusive-OR gate.*

*Exclusive-NOR*

An exclusive-NOR gate performs the opposite function to the exclusive-OR gate. Its symbol and truth table are given in Fig.8.7. It will be seen that output (logic 1) is obtained only when the inputs are the same or coincident. It is thus often referred to as a **coincident** or **equivalence** gate.

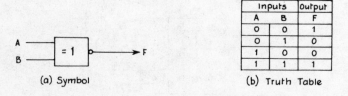

| Inputs | | Output |
|---|---|---|
| A | B | F |
| 0 | 0 | 1 |
| 0 | 1 | 0 |
| 1 | 0 | 0 |
| 1 | 1 | 1 |

(a) Symbol  (b) Truth Table

*Fig. 8.7 Exclusive NOR-gate.*

## Boolean Expressions

**Boolean** or **Switching** Algebra is a convenient way of representing the action of logic gates or systems and is simpler than truth tables, particularly when there is a large number of inputs. The symbols used relate to the fundamental logic gate operations of AND, OR and NOT:

| Logic Function | Boolean Expression. |
| --- | --- |

AND  $\quad F = A . B$ (where . means AND)

F (the output) is at logic 1 if the inputs A and B are both at logic 1.

OR  $\quad F = A + B$ (where + means OR)

F is at logic 1 if A or B is at logic 1.

NOT  $\quad F = \bar{A}$ (where $\bar{A}$ means the opposite of A)

F is at logic 1 when A is not at logic 1.

Thus for NAND and NOR we have:

NAND  $\quad F = \overline{A . B}$

NOR  $\quad F = \overline{A + B}$

and for EX–OR and EX–NOR we have:

EX–OR  $\quad F = (A . \bar{B}) + (\bar{A} . B)$

EX–NOR  $\quad F = (A . B) + (\bar{A} . \bar{B})$

## Combinational Logic

Logic gates are frequently used in combination with one another to produce complex logic systems and some further examples will be considered here.

*Example 1*

A self-service petrol pump will deliver petrol (F) if the grade selector is set at either 4-star (A) or 2-star (B) and the cashier switch is also operated. Draw a logic circuit that will implement this requirement giving the Boolean expression at the output and the truth table. A suitable logic circuit is given in Fig.8.8(a) requiring an OR gate G1 and an AND gate G2. The Boolean expression for the output is $F = (A + B) . C$.

In drawing up a truth table for combinational gates it is useful to tabulate the output of any intermediate gate(s) as shown in Fig.8.8(b). Here we see that the intermediate output (S) has been listed for the four combinations of the inputs A and B. The output at S thus provides one of the two inputs together with C for

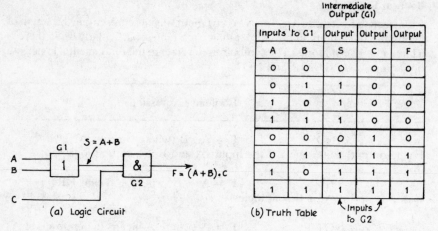

| Inputs to G1 | | Intermediate Output (G1) | | |
|---|---|---|---|---|
| | | Output | Output | Output |
| A | B | S | C | F |
| O | O | O | O | O |
| O | 1 | 1 | 0 | 0 |
| 1 | O | 1 | 0 | 0 |
| 1 | 1 | 1 | 0 | 0 |
| O | O | O | 1 | 0 |
| O | 1 | 1 | 1 | 1 |
| 1 | O | 1 | 1 | 1 |
| 1 | 1 | 1 | 1 | 1 |

(a) Logic Circuit    (b) Truth Table

Fig. 8.8 Logic circuit of petrol pump.

the AND gate G2. However, it will be noted that the four combinations of inputs A and B have been listed twice. This is because that for every combination of A and B, input C can be at logic 1 or logic 0. The listing thus gives a **full truth table**.

*Example 2*

A video cassette recorder is required to rewind the tape from right-to-left when the R button is operated but to fast-forward the tape from left-to-right when the F button is operated. The control system will need to provide two outputs, X to operate the rewind motor and Y to operate the fast-forward motor ('1' signifies output and '0' no output). A safeguard should be incorporated to protect the tape from breaking should the F and R buttons be operated simultaneously. Draw a logic circuit that will meet these requirements. Give the Boolean expression at each output and draw up a truth table.

A suitable logic circuit is given in Fig.8.9(a) using two AND gates and two inverter gates. It will be seen that output is obtained from G1 only when the R button is operated and the F button is not operated. Similarly an output is obtained from G2 only when the F button is operated and the R button is not operated. If both R and F buttons are operated simultaneously no output is obtained at X or Y so neither motor will turn. These conditions are confirmed by the truth table of Fig.8.9(b).

*Example 3*

If an industrial process control an alarm (F) is to be given if two liquid containers A and B are either both full (1) or both empty (0). No alarm is to be sounded if only one container is empty or only one container is full. Draw a logic circuit that will implement this requirement and produce a truth table. The required logic circuit is given in Fig.8.10(a). This is an equivalence or

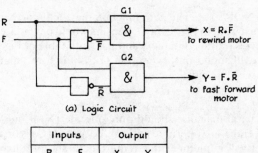

(a) Logic Circuit

| Inputs | | Output | |
|---|---|---|---|
| R | F | X | Y |
| O | O | O | O |
| O | 1 | O | 1 |
| 1 | O | 1 | O |
| 1 | 1 | O | O |

— Rewind motor operates
— Forward motor operates

(b) Truth Table

Fig. 8.9 Logic circuit for VCR motor control.

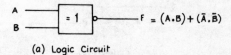

(a) Logic Circuit

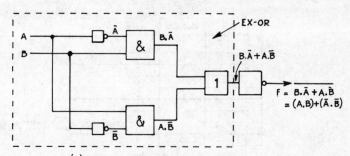

(b) EX-NOR made up from basic gates

| Inputs | | Output |
|---|---|---|
| A | B | F |
| O | O | 1 |
| O | 1 | O |
| 1 | O | O |
| 1 | 1 | 1 |

— Both containers empty
— Both containers full

(c) Truth Table

Fig. 8.10 Logic circuit for liquid container alarm.

Exclusive-NOR gate. The gate function may be implemented using basic logic gates as illustrated in Fig.8.10(b). Sensors would be required to detect the empty and full levels of the containers and the truth table of Fig.8.10(c) confirms that the alarm will sound only when the sensors indicate that both containers are empty or both are full.

*Example 4*

An electrical machine should only operate (X) if one and only one of two switches S and T are operated and a safety guard (Q) is in position. Draw a logic circuit that will implement this requirement and produce a truth table.

The logic solution is an EX-OR gate and an AND gate connected as shown in Fig.8.11(a). The truth table is given in Fig.8.11(b); note again the use of an intermediate output in forming the full truth table. The use of the EX-OR gate permits one and only one switch to operate the machine.

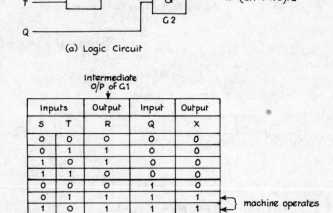

(a) Logic Circuit

Intermediate O/P of G1

| Inputs | | Output | Input | Output |
|---|---|---|---|---|
| S | T | R | Q | X |
| 0 | 0 | 0 | 0 | 0 |
| 0 | 1 | 1 | 0 | 0 |
| 1 | 0 | 1 | 0 | 0 |
| 1 | 1 | 0 | 0 | 0 |
| 0 | 0 | 0 | 1 | 0 |
| 0 | 1 | 1 | 1 | 1 |
| 1 | 0 | 1 | 1 | 1 |
| 1 | 1 | 0 | 1 | 0 |

} machine operates

Inputs to G2

Fig. 8.11 Logic circuit for machine control.

*Example 5*

Using NAND gates only, draw logic diagrams to perform the OR and NOR operations. Fig.8.12(a) and (b) show how 2-input NAND gates may be arranged to perform the OR and NOR operations. Note that when the 2 inputs of the NAND gate are commoned it behaves as a NOT gate.

There are many attractions in standardising with a single type of gate (NAND or NOR), particularly with large-scale integration. The NAND and NOR gates

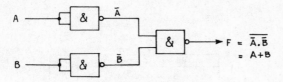

(a) Use of NAND gates to produce OR function

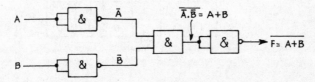

(b) Use of NAND gates to produce NOR function

Fig. 8.12 NAND gate standardisation.

are 'universal', i.e. any other type of gate may be synthesised by using just NAND or NOR alone. The diagram of Fig.8.13 shows how the Exclusive-OR function may be implemented using just NAND gates.

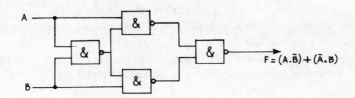

Fig. 8.13 Use of NAND gates to produce Exclusive-OR.

*Example 6*

Draw a truth table for the logic diagram of Fig.8.14 including the intermediate outputs X, Y and Z. From the truth table deduce a simpler form for the arrangement. The truth table is given in Fig.8.15. It will be seen that the arrangement effectively performs the NAND function, thus the logic diagram may be replaced by a 2-input NAND gate.

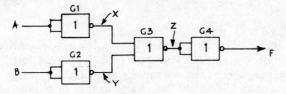

Fig. 8.14

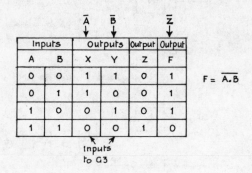

| Inputs | | Outputs | | Output | Output |
|---|---|---|---|---|---|
| A | B | X | Y | Z | F |
| 0 | 0 | 1 | 1 | 0 | 1 |
| 0 | 1 | 1 | 0 | 0 | 1 |
| 1 | 0 | 0 | 1 | 0 | 1 |
| 1 | 1 | 0 | 0 | 1 | 0 |

$F = \overline{A.B}$

Inputs to G3

*Fig. 8.15  Truth table for Fig. 8.14.*

## SEQUENTIAL LOGIC CIRCUITS

In sequential logic circuits, the output depends upon binary signals which have already been applied over some previous period of time. A sequential system must therefore possess some 'storage' or 'memory' device. It should be noted that combinational logic elements are also involved in sequential systems.

### Bistable Element

A bistable is a simple memory or storage element and has two stable states. Once the device has been put into one state it will remain in that state until a signal is applied to change it to its new second stable state. If the power supply is disrupted, the state to which the bistable will return when power is restored is indeterminate.

The basic principle of the bistable oscillator was dealt with in Chapter 7 and it will now be shown how the bistable can be formed from logic gates.

### S-R Bistable

The SET-RESET bistable is the fundamental bistable element and its symbol is given in Fig.8.16(a). It has two inputs SET (S) and RESET (R) and two outputs Q and $\bar{Q}$, where $\bar{Q}$ means the opposite of Q, i.e. if Q = 0 then $\bar{Q} = 1$ and vice-versa. The S-R bistable may be formed from two cross-coupled NOR gates as shown in Fig.8.16(b).

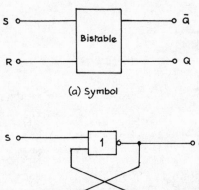

(a) Symbol

(b) S-R Bistable using NOR gates

Fig. 8.16 S-R bistable.

## Operation

**S = 0, R = 0.**   When power is applied to the bistable, the Q and $\bar{Q}$ outputs will take up opposite logic states. Suppose that initially that $Q = 0$ and $\bar{Q} = 1$ and that $S = 0$, $R = 0$, as shown in Fig.8.17.

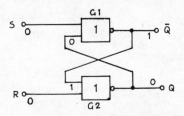

Fig. 8.17 S = 0, R = 0 (no change-store).

It will be seen from the logic states on the diagram that this is a stable operating condition. Since $Q = 0$, both inputs to G1 will be at logic 0 and its output at logic 1. Since $\bar{Q} = 1$, one of the inputs to G2 will be at logic 1 thus its output must be at logic 0. It follows that Q holds $\bar{Q}$ at logic 1 and $\bar{Q}$ holds Q at logic 0.

**S = 1, R = 0.**   Suppose now that input S is taken to the logic 1 state with input R remaining at logic 0. The initial state is therefore as in Fig.8.18(a). The application of a logic 1 to G1 from the S input will cause the $\bar{Q}$ output to change

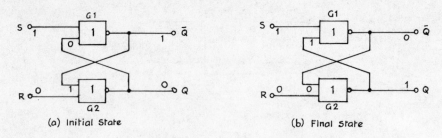

*(a) Initial State*      *(b) Final State*

*Fig. 8.18  S = 1, R = 0 (set Q to logic 1).*

state to logic 0 as shown in Fig.8.18(b). This results in both inputs to G2 becoming logic 0 causing the Q output to change state to logic 1. If the S line is now returned to logic 0 it will be seen that the Q output remains at logic 1 ($\bar{Q} = 0$). The logic 1 need only be applied to the S input momentarily since the feedback between gates causes the bistable to memorise the instruction. It should be noted that a further logic 1 applied to S has no effect once Q = 1.

**S = 0, R = 1.** Supposing now that after setting the Q output to logic 1, the R input is placed at logic 1 with the S input held at logic 0. The initial state is then as in Fig.8.19(a). The application of the logic 1 to G2 from the R input will cause the Q output to change state to logic 0. This action results in both inputs to G1 assuming the logic 0 state and for the $\bar{Q}$ output to change to logic 1, see Fig.8.19(b). Again the logic 1 need only be applied momentarily. If the R input is returned to logic 0 it will be seen that the Q output will remain at logic 0.

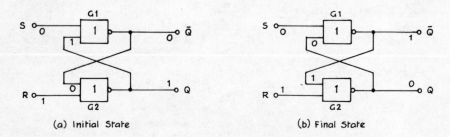

*(a) Initial State*      *(b) Final State*

*Fig. 8.19  S = 0, R = 1 (reset Q output to logic 0).*

**S = 1, R = 1.** If the S and R inputs are simultaneously set at logic 1, see Fig.8.20, then both the Q and $\bar{Q}$ outputs will be at logic 0. If then the S and R inputs are returned to logic 0 the output will be indeterminate depending upon the relative switching speed of the two gates, i.e. if G2 is the faster gate the Q output will be set at logic 1 but if G1 is the faster the $\bar{Q}$ output will go to logic 1. Because of the uncertainty of the output state, the condition S = R = 1 must be avoided.

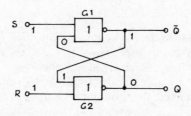

Fig. 8.20 S = 1, R = 1 (output state indeterminate).

As with other logic devices, we may draw a truth table for the S-R bistable which summarises its operation, see Fig.8.21. In the table, the column $Q_n$ represents the state of the Q output **prior to** the application of the inputs S and R and $Q_{n+1}$ represents the state of the Q output **after** the application of the listed inputs at S and R.

| S | R | $Q_n$ | $Q_{n+1}$ | |
|---|---|---|---|---|
| 0 | 0 | 0 | 0 | } No change (store condition) |
| 0 | 0 | 1 | 1 | |
| 1 | 0 | 0 | 1 | } Q output SET at Logic 1 |
| 1 | 0 | 1 | 1 | |
| 0 | 1 | 0 | 0 | } Q output RESET at Logic 0 |
| 0 | 1 | 1 | 0 | |
| 1 | 1 | 0 | ? | } Indeterminate |
| 1 | 1 | 1 | ? | output state |

Fig. 8.21 Truth table for S-R bistable.

The S-R bistable may be constructed from NAND gates as shown in Fig.8.22.

Typical applications for S-R bistables include temporary stores for binary information and switch debouncing.

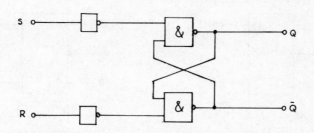

Fig. 8.22 S-R bistable formed from cross-coupled NAND gates.

## Switch Debouncer

When mechanical switches are used to provide logic inputs to digital circuits, 'contact bounce' can lead to problems as the contacts are closed. The effect of the bounce is to produce several pulses when only **one** should have been produced. The S-R bistable can be used to eliminate multiple output, see Fig.8.23. On the first make the Q output is set to logic 1, the remaining bounces

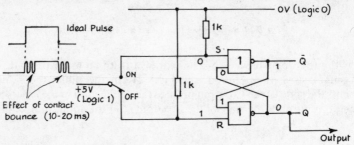

Fig. 8.23 Use of S-R bistable as switch debouncer.

have no further effect. In the 'off' position of the switch, the logic states are as shown in the diagram. On the first make, the Q output will be set at logic 1 ($\bar{Q}$ at logic 0) since the S input will take up the logic 1 state and the R input the logic 0 state. As the switch contacts break, both the S and R inputs will assume the logic 0 state and thus the output will remain unchanged. On the next make when S = 1 and R = 0 the Q output will remain in the logic 1 state since the bistable was previously 'set'.

## Clocked S-R Bistable

It is often desirable to control synchronously all operations in a digital logic system and this may be achieved using 'synchronising' or 'clock' pulses. By using clock pulses generated by a stable oscillator it is possible to 'gate' or 'clock' the logic levels at the S and R inputs into the bistable at some precise instant in time. The clocking may be implemented on the leading edges of the clock pulses as in Fig.8.24(a) or on the trailing edges as in Fig.8.24(b).

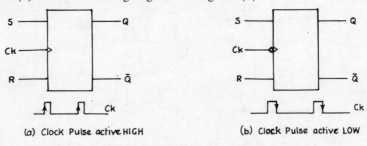

Fig. 8.24 Clocked S-R bistable symbols.

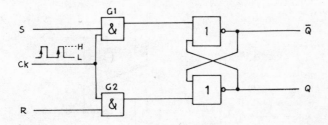

Fig. 8.25 Method of producing a clocked S-R bistable (clock pulse active HIGH).

Fig.8.25 shows one way of producing a clocked S-R bistable, clock pulse active HIGH. When the clock pulse is LOW, the output from the AND gates G1 and G2 will be LOW. Changes in the logic conditions at the S and R inputs will have no effect and the bistable will be in the 'store' mode. When the clock pulse input goes HIGH on its leading edge, the outputs of G1 and G2 will correspond to the S and R inputs respectively and the bistable may change state accordingly. This bistable may change state therefore only when the clock input is HIGH. The complete truth table for the clocked S-R bistable is given in Fig.8.26.

| Inputs | | | Output | |
|---|---|---|---|---|
| Clock | S | R | $Q_n$ | $Q_{n+1}$ |
| 0 | 0 | 0 | 0 | 0 |
| 0 | 0 | 0 | 1 | 1 |
| 0 | 0 | 1 | 0 | 0 |
| 0 | 0 | 1 | 1 | 1 |
| 0 | 1 | 0 | 0 | 0 |
| 0 | 1 | 0 | 1 | 1 |
| 0 | 1 | 1 | 0 | 0 |
| 0 | 1 | 1 | 1 | 1 |
| 1 | 0 | 0 | 0 | 0 |
| 1 | 0 | 0 | 1 | 1 |
| 1 | 0 | 1 | 0 | 0 |
| 1 | 0 | 1 | 1 | 0 |
| 1 | 1 | 0 | 0 | 1 |
| 1 | 1 | 0 | 1 | 1 |
| 1 | 1 | 1 | 0 | ? |
| 1 | 1 | 1 | 1 | ? |

$Q_n$ = state prior to application of inputs

$Q_{n+1}$ = state after application of inputs

} Indeterminate

Fig. 8.26 Truth table for clocked bistable.

Additional inputs $S_D$ and $R_D$ may be provided (direct SET and direct RESET) as illustrated in Fig.8.27. These inputs may be used to SET the Q output of the bistable to logic 1 or RESET it to logic 0 independently of the state of the S, R and Ck inputs. The $S_D$ and $R_D$ inputs are 'forcing' inputs and are sometimes called the PRESET and CLEAR inputs respectively.

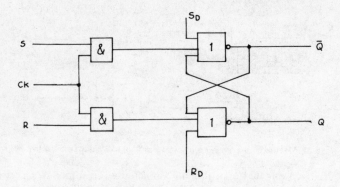

Fig. 8.27 Use of $S_D$ and $R_D$ (Preset & Clear) inputs.

## The J-K Bistable

The J-K bistable is essentially an S-R bistable with additional logic circuitry to eliminate the indeterminate output that occurs when S = R = 1. One possible circuit arrangement is given in Fig.8.28. It is a clocked bistable and it is provided with two input terminals (J and K); these are 'preparatory' inputs, i.e. they prepare the circuit for a change under the control of the clock input. Feedback from the Q and Q̄ outputs and the two AND gates provide a **clock steering function** and the condition J = K = 1 is now permissible resulting in a predictable output. Direct SET and Direct RESET (PR and CLR) may also be provided, see logic circuit symbol.

It will be seen that the Q and Q̄ outputs as well as the preparatory inputs J and K determine which input of the basic S-R bistable receives the clock pulse. For

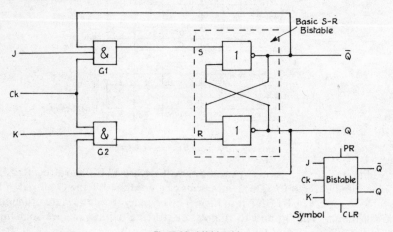

Fig. 8.28 J-K bistable.

example, consider that initially Q = 0 and $\bar{Q}$ = 1, a logic 1 is applied to J and a logic 0 to the K input. Because of the feedback, then when a clock pulse is applied (active high) the upper AND gate G1 will be operative and the lower AND gate G2 inoperative. Thus the logic states of J and K are transferred to the S and R inputs of the basic S-R bistable, setting the Q output to logic 1 and the $\bar{Q}$ output to logic 0.

If now a logic 1 is applied to the K input and a logic 0 to the J input, then on the next clock pulse G2 will be made operative and G1 inoperative. As a result the Q output will be reset to logic 0 and the $\bar{Q}$ output to logic 1.

In principle, the results set out in the truth table of Fig.8.29 will apply. The

| J | K | $Q_n$<br>Before clock pulse | $Q_{n+1}$<br>After clock pulse |
|---|---|---|---|
| 0 | 0 | 0 | 0 |
| 0 | 0 | 1 | 1 |
| 0 | 1 | 0 | 0 |
| 0 | 1 | 1 | 0 |
| 1 | 0 | 0 | 1 |
| 1 | 0 | 1 | 1 |
| 1 | 1 | 0 | 1 |
| 1 | 1 | 1 | 0 |

*Fig. 8.29 Truth table for J-K bistable.*

table illustrates the versatility of the J-K bistable which may be summarised as follows:

(1) If J = K = 0 there will be no change in the output state when a clock pulse is applied (the bistable is then in the HOLD condition).

(2) If J = 1, K = 0 the Q output is placed in the 1 state and the $\bar{Q}$ in the 0 state when a clock pulse is applied (the bistable is then in the SET condition).

(3) If J = 0, K = 1 the Q output is placed in the 0 state and the $\bar{Q}$ in the 1 state when a clock pulse is applied (the bistable is then in the RESET condition).

(4) If J = K = 1, the Q and $\bar{Q}$ outputs reverse states on the receipt of each clock pulse (the bistable is then in the TOGGLING mode).

It will therefore be appreciated that there are more modes available with this type of bistable due to the use of the preparatory inputs J and K.

*Race-Around Condition*

If the Q and $\bar{Q}$ outputs of the J-K bistable change state before the end of the clock pulse, the input conditions to the AND steering gates of Fig.8.28 will change. The effect is that Q oscillates between 1 and 0 for the duration of the clock pulse, see Fig.8.30, and at the end of the clock pulse period the output is indeterminate.

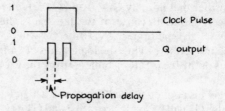

Fig. 8.30 Race-around condition.

The condition can be avoided if the clock pulse period is short compared with the 'propagation delay' (switching time) of the bistable. Because this requirement is rarely met in practice with modern high-speed integrated circuits, the race-around condition led to the development of the Master–Slave bistable.

### Master–Slave J-K Bistable

This type of J-K bistable does not suffer from the 'race-around' condition and the basic idea of its operation is illustrated by Fig. 8.31.

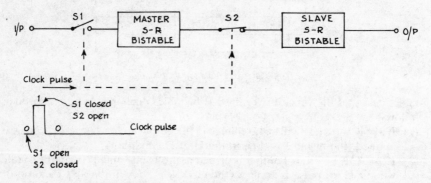

Fig. 8.31 Basic idea of Master–Slave J-K bistable.

There are now two bistables, the 'Master' and the 'Slave', with the Slave driven from the Master. The switches S1 and S2 are arranged so that when S1 is closed, S2 is open and vice versa. On the leading edge of the clock pulse S1 is closed and the input is clocked into the output of the Master. At this time the Slave bistable will not change since switch S2 is open. When the clock pulse changes state once more on its trailing edge, S1 opens disconnecting the input from the Master. At the same time, since S2 is closed the output of the Slave is made to follow the output of the Master. Because S1 is open, any feedback applied in an actual circuit cannot affect the output of the Master.

It will now be shown how this basic principle is applied to the J-K bistable, a circuit arrangement of which is given in Fig.8.32. As with the basic J-K bistable,

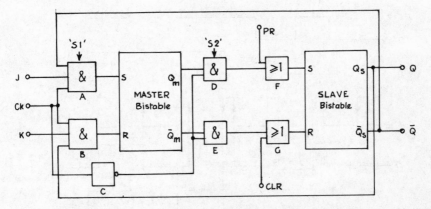

Fig. 8.32 Circuit arrangement for J-K Master–Slave bistable.

feedback is applied from the output (of the SLAVE) to the clock steering gates A and B. This configuration eliminates the 'race-around' condition and a predictable output results for the condition J = K = 1.

*Operation*

The AND gates A and B effectively replace 'S1' of Fig.8.31 and the AND gates D and E replace 'S2'. When the clock pulse goes to logic 1 on its leading edge, the Master is enabled via gates A and B but due to the inverter gate C, the Slave is disabled. The following initial conditions will be assumed:

$Qs = 0$ , $Qm = 0$ , $J = K = 1$, Clock pulse = 0 and PR = 0
$\bar{Q}s = 1$ , $\bar{Q}m = 1$                                    CLR = 0

(1) When the clock pulse changes from 0 to 1 on its leading edge, the output of gate A will go to logic 1 but the gate B output will be in the logic 0 state. Since gate A and B outputs are the S and R inputs of the Master, the Qm output of the Master will be set at logic 1 ($\bar{Q}m = 0$). At this time the Slave is disabled as gates D and E are inoperative.

(2) When the clock pulse changes from logic 1 to logic 0 on its trailing edge, gates A and B are disabled, but gates D and E are enabled. Thus the Qm and $\bar{Q}m$ outputs of the Master become the inputs of the Slave, i.e. Qs goes to logic 1 and $\bar{Q}s$ goes to logic 0. The change in the output states of the Slave has no effect on the Master since gates A and B are inoperative.

(3) On the leading edge of the next clock pulse when it goes from logic 0 to logic 1, the output of gate A is set at 0 and the output of gate B to logic 1. This resets the Master outputs to Qm = 0 and $\bar{Q}m = 1$. On the trailing edge of the clock pulse, the Slave follows the Master and the Slave outputs are reset to Qs = 0, $\bar{Q}s = 1$.

When the Master–Slave Bistable is operated in this way it is said to be a

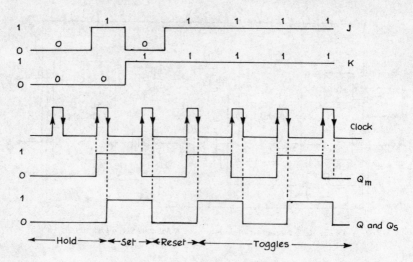

Fig. 8.33 Waveforms illustrating operation of Master–Slave J-K bistable.

'toggle' bistable (T-Bistable). Waveforms summarising the modes of operation of the Master–Slave Bistable are illustrated in Fig.8.33.

The J-K bistable is a very important and flexible logic device and forms the basis of counting circuits which are considered in the next section. Fig.8.34 shows how the T-bistable is constructed by tying the J and K terminals together (now called the T-terminal). When T is at logic 0, the state of the bistable will not change when the clock input goes to logic 1. However when T is at logic 1, the bistable will change state each time the clock input goes to logic 1 (high).

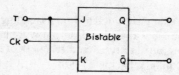

Fig. 8.34 T-bistable.

# COUNTERS

### Asynchronous Binary Counter

A common method of producing a binary counter is to use a number of Master–Slave J-K bistables wired as 'toggle' bistables as shown in Fig.8.35. The pulses to be counted are applied to the clock input of bistable A and it will be

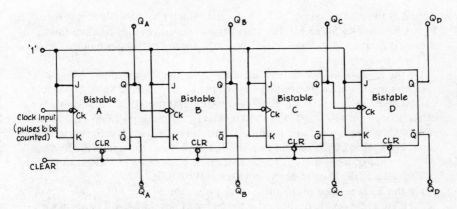

*Fig. 8.35 Asynchronous binary counter (Modulo-16).*

seen that the output of bistable A provides the clock input to bistable B; the output of bistable B provides the clock input to bistable C, and so on. Negative-edge triggering is used, i.e. the 'slave' output of any bistable changes state on the negative-going edge of its clock input. A common 'clear' line (clear LOW) is used to reset all Q outputs LOW. During counting the 'clear' line is held in the HIGH state. With this particular circuit both Q and $\bar{Q}$ outputs are made available, but this is not always the case.

### Operation

Assume that the counter has been reset by taking the 'clear' line LOW, i.e. $Q_A - Q_E$ outputs will all be LOW (logic 0). The waveforms given in Fig.8.36 depict the operation during counting.

On the leading edge of the first clock pulse the 'Master' of bistable A will be set at logic 1 and on the trailing edge, the logic 1 from the 'Master' will be transferred to the output of the 'Slave', i.e. $Q_A$ will be set at logic 1. Since the Q

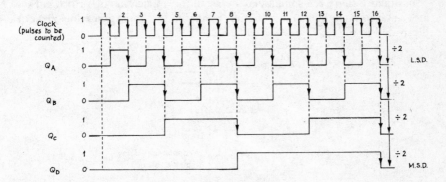

*Fig. 8.36 Waveforms showing operation of asynchronous counter of Fig. 8.35.*

output is the clock input of bistable B, the 'Master' of bistable B will be set at logic 1 (but not its 'Slave') at this time. Thus at the end of the first clock pulse, the binary output of the counter will be 0001 (note that $Q_A$ provides the least significant digit).

On the leading edge of the second clock pulse the Master of bistable A will be reset to logic 0 and on the trailing edge of the clock pulse the Slave of bistable A will be reset to logic 0. It will be remembered that the Master of bistable B has already been set at logic 1, thus when the $Q_A$ output goes LOW, the Slave of bistable B will be set to logic 1. Since the output of bistable B is connected to the input of bistable C, the Master of bistable C will be set to logic 1 (the Slave of C will not change until the output of bistable B goes LOW). Thus at the end of clock pulse 2, the output of the counter will be 0010.

On the leading edge of the third clock pulse, the Master of bistable A will be set at logic 1 and on the trailing edge the Slave of bistable A will be set at logic 1. $Q_A$ will change from logic 0 to logic 1 and the Master of bistable B will be reset to logic 0. At the end of clock pulse 3, the output of the counter will be 0011.

This procedure continues and at the end of 15 clock pulses the counter output will be 1111. Upon receipt of clock pulse 16, the counter will be reset to 0000.

### Modulus

Any number of bistables can be cascaded in this way to form an n-bit counter. The arrangement of Fig.8.35 is referred to as a 4-bit counter since there are 4 binary bits or digits in the coded output. A 5-bit binary counter will count up to decimal 31 and reset on clock pulse 32, but requires 5 bistables of course. The modulus of a binary number system is the decimal equivalent which that number of bits or digits can represent, e.g.

3-bit binary counter = $2^3$ = 8 — Modulo-8
4-bit binary counter = $2^4$ = 16 — Modulo-16
8-bit binary counter = $2^8$ = 256 — Modulo-256

### Propagation Delay

In practice there is a small delay between the application of a clock pulse to any stage of a counter and its output assuming the appropriate logic level as illustrated in Fig.8.37.

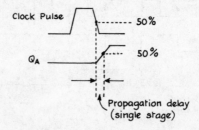

Fig. 8.37 Maximum counting rate limited by propagation delay.

With an asynchronous counter or 'ripple' counter, the bistables do not change state at the same instant, causing the propagation delays to become accumulative. For example, when the counter output is 0111 then on the eighth clock pulse all the bistables change state. However $Q_D$ cannot change until $Q_C$ changes, $Q_C$ cannot change until $Q_B$ changes and $Q_B$ cannot change until $Q_A$ changes. Thus, if the propagation delay of each stage is 20 ns, a time delay of 80 ns will elapse before the counter output finally settles to 1000. Because the change of state appears to 'ripple' through the counter, the asynchronous counter is also known as a **ripple-counter**.

### Bidirectional Binary Counter

The asynchronous binary counter previously described is a FORWARD counter or an UP counter since it counts 'up' from zero. It is often required to be able to count 'down' from a pre-determined value to zero. A counter operating in this mode is referred to as a DOWN counter or REVERSE counter.

From the truth table given in Fig.8.38 it will be seen that the **total** value stored in both Q and Q̄ at the end of any clock pulse is constant and equal to decimal 15.

| Clock Pulse No. | Output States | | | | | | | | |
|---|---|---|---|---|---|---|---|---|---|
| | $Q_D$ | $Q_C$ | $Q_B$ | $Q_A$ | | | $\bar{Q}_D$ | $\bar{Q}_C$ | $\bar{Q}_B$ | $\bar{Q}_A$ |
| 0 (16) | 0 | 0 | 0 | 0 | | | 1 | 1 | 1 | 1 |
| 1 | 0 | 0 | 0 | 1 | | | 1 | 1 | 1 | 0 |
| 2 | 0 | 0 | 1 | 0 | | | 1 | 1 | 0 | 1 |
| 3 | 0 | 0 | 1 | 1 | Count 'Up' | Count 'Down' | 1 | 1 | 0 | 0 |
| 4 | 0 | 1 | 0 | 0 | | | 1 | 0 | 1 | 1 |
| 5 | 0 | 1 | 0 | 1 | | | 1 | 0 | 1 | 0 |
| 6 | 0 | 1 | 1 | 0 | | | 1 | 0 | 0 | 1 |
| 7 | 0 | 1 | 1 | 1 | | | 1 | 0 | 0 | 0 |
| 8 | 1 | 0 | 0 | 0 | | | 0 | 1 | 1 | 1 |
| 9 | 1 | 0 | 0 | 1 | | | 0 | 1 | 1 | 0 |
| 10 | 1 | 0 | 1 | 0 | | | 0 | 1 | 0 | 1 |
| 11 | 1 | 0 | 1 | 1 | | | 0 | 1 | 0 | 0 |
| 12 | 1 | 1 | 0 | 0 | | | 0 | 0 | 1 | 1 |
| 13 | 1 | 1 | 0 | 1 | | | 0 | 0 | 1 | 0 |
| 14 | 1 | 1 | 1 | 0 | | | 0 | 0 | 0 | 1 |
| 15 | 1 | 1 | 1 | 1 | | | 0 | 0 | 0 | 0 |

*Fig. 8.38 Truth table for binary counter of Fig. 8.35.*

Clearly the previous counter will count 'down' if the Q̄ outputs are used. By using suitable combinational logic a **Bi-Directional** counter may be constructed. One possible circuit arrangement is given in Fig.8.39.

To provide the reversible function with a single counter it is necessary to introduce a control line as shown. When the control line is set at logic 1 the upper set of AND gates are enabled but the lower set of AND gates are disabled (by the inverter gate) and the counter is in the UP count mode. When the control

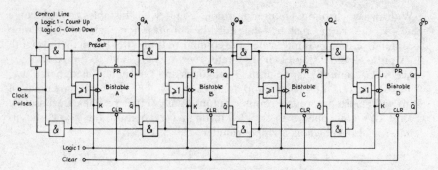

*Fig. 8.39 Reversible (up-down) counter.*

line is set at logic 0, the upper set of AND gates are disabled but the lower set are enabled; the counter is then in the DOWN count mode.

### Count UP Mode

The control line will be held at logic 1 and if the clear line is taken LOW the counter outputs will all be reset at logic 0. The clear line is then taken HIGH. When the clock pulses (the pulses to be counted) are applied at the counter input, the counter will count up as previously described for Fig.8.35 with the upper set of AND gates connecting the Q output of each bistable to the clock input of the following bistable via an OR gate. After 15 clock pulses the counter will register 1111 at its outputs.

### Count DOWN Mode

If now the control line is taken to logic 0, the counter will count down from 1111. On the leading edge of the first clock pulse, the Ck input of bistable A will go HIGH and the Master of bistable A will be reset. The Ck input of bistable B will remain LOW since $\bar{Q}_A$ is logic 0. On the trailing edge of the clock pulse, the Slave of bistable A will be reset to logic 0 and $\bar{Q}_A$ will change to logic 1. Throughout this interval, the Ck input of bistable B has remained LOW. The counter output is now 1110.

On the leading edge of the second clock pulse, the Ck input of bistable A will go HIGH and the Master will be set. At the same time the Ck input of bistable B will go HIGH and its Master will be reset to logic 0. On the trailing edge of the clock pulse, the Slave of bistable A will be set to logic 1 and the Slave of bistable B will be reset to logic 0. The counter output is now 1101.

This operation continues down the counter with the counter output being decremented by one for each clock pulse input.

### Decade Counter

It is often desirable to be able to count to a base N which is not a power of 2. For example, to count to a base of 10 is a common requirement as this is the basis of the denary number system of which we are more familiar.

For a decade or divide-by-10 counter we require a chain of n bistables such that n is the smallest number for which $2^n > N$, e.g. for $N = 10$ (decade counter), $n = 4$. A feedback gate is then added so that on a count of 10 all outputs are reset to logic 0. Inspection of the truth table for a binary counter shows that at a count of 10 the output states are:

$$Q_D = 1, Q_C = 0, Q_B = 1, \text{ and } Q_A = 0 \ (1010).$$

However, for a decade counter we require **all** of the outputs to be in the logic 0 state at a count of 10, i.e. the $Q_D$ and $Q_B$ outputs must be reset back to logic 0 at a count of 10. One way of achieving this is as shown in Fig.8.40.

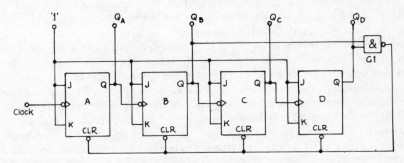

Fig. 8.40 Asynchronous decade counter.

The four bistable stages are wired as toggle bistables with the pulses to be counted applied to the first stage. When a count of 1010 is reached, the output from the NAND gate G1 goes LOW which is applied to the 'clear' inputs of all stages, resetting all outputs to logic 0. The timing waveforms for the counter are shown in Fig.8.41. Note that on the 10th input pulse the counter outputs are all at

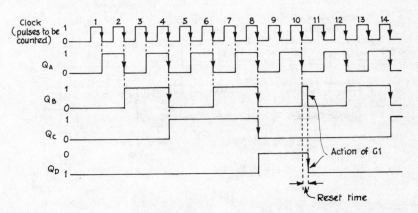

Fig. 8.41 Waveforms showing operation of decade counter.

logic 0; this is providing that the propagation delay (reset time) of the CLEAR inputs is short (typically 50 ns for TTL devices). Unequal reset times from the CLEAR to the output of individual bistables may cause unreliable operation, e.g. if bistable B has a shorter reset time than bistable D, then bistable B input to G1 will go LOW before bistable D input, with the result that bistable D will not clear. It will be seen that the counting sequence recommences on the 11th input pulse.

## DISPLAY DRIVER/DECODERS

7-segment displays using l.e.d. and liquid crystal segments were described in Volume 2 of this series. It will now be shown how the binary signal output from a digital circuit may be decoded and hence used to drive the display segments. Consider the arrangement in Fig.8.42 where decade counters are used to operate a 7-segment display providing a 3-digit decimal read-out. Three separate counters are employed arranged as the 'Units', 'Tens' and 'Hundreds' decade counters thus providing binary-coded decimal (b.c.d.) outputs. The pulses to be counted are applied to the clock input of the 'units' counter, the $Q_D$ output of which is used as the clock input of the 'tens' decade counter. The $Q_D$ output of the 'tens' counter is then used as the clock input of the 'hundreds' decade counter; this idea may be extended to increase the number of digits used in the display.

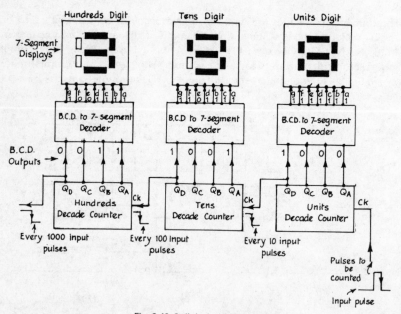

Fig. 8.42 3-digit decimal read out.

It will be appreciated, of course, that although a decade counter counts input pulses from 0 to 9 and then resets on the 10th pulse, the actual read-out from the bistable stages is in **binary**. However since only the binary digits equivalent to 0–9 in decimal are involved in the output from each decade counter, the output is said to be in binary-coded decimal. To operate the 7-segment displays the counter outputs must be 'decoded' using 4–7 line decoders. The diagram shows the decoder output states after 398 pulses have been applied to the input of the 'units' counter.

### Decoders

A **decoder** is a combinational logic circuit that is used to decode or translate a number of input binary lines into a number of output lines where for a given binary input, **one** of the output lines goes HIGH all other lines being LOW or vice versa.

A simple arrangement is the **2-to-4 line decoder**, a logic circuit of which is given in Fig.8.43. The logic states shown on the diagram illustrate the operation for a binary input of 11. In this arrangement a LOW (logic 0) is obtained on only one of the output lines for each binary code on the input, see the truth table of Fig.8.43(b). Thus the presence of a logic 0 on a particular output line identifies

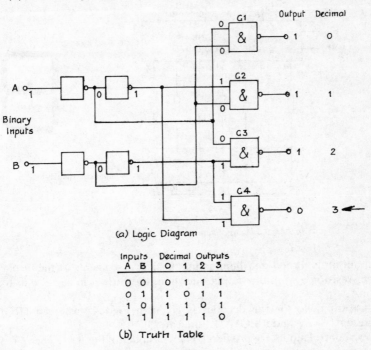

(a) Logic Diagram

| Inputs | | Decimal Outputs | | | |
|---|---|---|---|---|---|
| A | B | 0 | 1 | 2 | 3 |
| 0 | 0 | 0 | 1 | 1 | 1 |
| 0 | 1 | 1 | 0 | 1 | 1 |
| 1 | 0 | 1 | 1 | 0 | 1 |
| 1 | 1 | 1 | 1 | 1 | 0 |

(b) Truth Table

Fig. 8.43 2-4 line decoder.

the binary code on the input; this is the **decoding** process, i.e. 2-line binary to 4-line decimal. The idea of Fig.8.43(a) may be extended to provide 8 output lines from 3 binary input lines or 16 output lines from 4 binary input lines, etc. but at the expense in the complexity in the number of logic gates used in the combination.

### B.C.D. -to- 7-Segment Decoder

A B.C.D. -to- 7-segment decoder has 4 input lines and 7 output lines and is formed from a combinational logic circuit. We need not concern ourselves with the actual logic circuit as the decoder is normally available as an integral i.c. package containing the combinational logic circuit together with driver output stages to operate the various segments. For each 4-bit b.c.d. code applied at the input the 7 output lines take up logic states (HIGHS or LOWS) so that the desired numeral is displayed.

The arrangement of Fig.8.44 illustrates how the decoder/driver is used to operate a 7-segment l.e.d. display (common-cathode type). When any of the output lines a–g of the decoder are HIGH (+ 5V), the l.e.d. array connected to

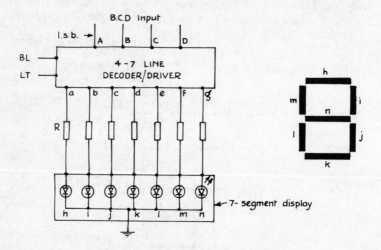

*Fig. 8.44 B.C.D. -to- 7-segment decoder/driver for l.e.d. display.*

that output is turned on, illuminating particular segments of the display. The series resistors (R) in the output lines limit the current in each l.e.d. to about 20 mA.

A truth table for the decoder is given in Fig.8.45 where a HIGH (H) represents + 5V and a LOW (L) indicates 0 V.

Two control inputs are provided on the decoder of Fig.8.44. The L.T. (Lamp Test) terminal is used to check that all outputs go HIGH when the L.T. terminal

| Inputs | | | | Outputs | | | | | | |
|---|---|---|---|---|---|---|---|---|---|---|
| D | C | B | A | a | b | c | d | e | f | g |
| L | L | L | L | H | H | H | H | H | H | L |
| L | L | L | H | L | H | H | L | L | L | L |
| L | L | H | L | H | H | L | H | H | L | H |
| L | L | H | H | H | H | H | H | L | L | H |
| L | H | L | L | L | H | H | L | L | H | H |
| L | H | L | H | H | L | H | H | L | H | H |
| L | H | H | L | H | L | H | H | H | H | H |
| L | H | H | H | H | H | H | L | L | H | L |
| H | L | L | L | H | H | H | H | H | H | H |
| H | L | L | H | H | H | H | H | L | H | H |

Fig. 8.45 Truth table for B.C.D. -to- 7-segment decoder.

is taken LOW. Under this condition all segments of the display should be illuminated, thus displaying numeral 8. The L.T. terminal is a master input and overrides any other input to the decoder. The BL (Blanking) terminal input is used to place the decoder outputs a–g in the LOW state irregardless of the state of the b.c.d. input and thus turns off the display. This control input is normally used to turn off leading zeros in a numeric display of digits, such as in a pocket calculator, to make the display easier to read and to reduce consumption.

## Liquid Crystal Display

Liquid crystal displays are also 7-segment devices and will thus require a 4–7 line decoder for displaying numerals 0 to 9. When a voltage is applied between the Backplate (BP) and the segment, that particular segment is opaque (black). On the other hand, if no p.d. exists between a segment and the backplate, that segment is transparent.

To prevent chemical deterioration and hence increase life expectancy, the display is normally driven by a square-wave applied to the Phase (PH) input of the decoder, see Fig.8.46. The diagram shows the logic states for an input that will result in the display of numeral 3. It will be noted that for this condition the decoder outputs g and f are logic 1 and logic 0 respectively. These states are also shown in diagrams (a) and (b) of Fig.8.47 together with the square-wave input and output of the respective EX-OR gate. It will be seen that the square-waves fed to the backplate and segment n are in antiphase, i.e. a p.d. exists between the backplate and segment n, whereas the square-waves fed to the backplate and segment m are in phase, i.e. no p.d. exists between the backplate and segment m. Thus segment n will be opaque and segment m transparent.

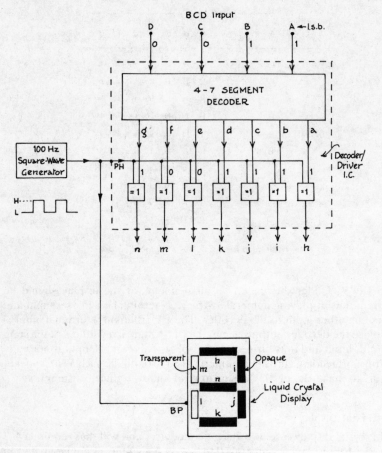

Fig. 8.46 Liquid crystal decoder/driver i.c.

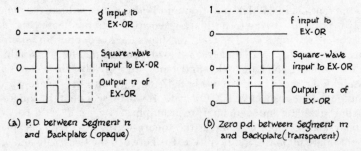

(a) P.D between Segment n and Backplate (opaque)

(b) Zero p.d. between Segment m and Backplate (transparent)

Fig. 8.47 Waveforms explaining operation of liquid crystal display.

## QUESTIONS ON CHAPTER EIGHT

(1)   An example of negative logic is:
    (a)   0V (logic 1), −5V (logic 0)
    (b)   +5V (logic 1), 0V (logic 0)
    (c)   −1V (logic 1), −5V (logic 0)
    (d)   −5V (logic 1), 0V (logic 0).

(2)   The truth table shown under gives the logic states for a:
    (a)   NAND gate
    (b)   EX-NOR gate
    (c)   EX-OR gate
    (d)   NOR gate.

| A | B | F |
|---|---|---|
| 0 | 0 | 1 |
| 0 | 1 | 0 |
| 1 | 0 | 0 |
| 1 | 1 | 1 |

(3)   The Boolean expression $F = \overline{A + B}$ represents a:
    (a)   NAND gate
    (b)   EX-OR gate
    (c)   NOR gate
    (d)   EX-NOR gate.

(4)   When $S = 1$ and $R = 0$, the output states of an S-R bistable are normally:
    (a)   $\bar{Q} = 1, Q = 0$
    (b)   $\bar{Q} = 0, Q = 1$
    (c)   $Q = 1, \bar{Q} = 1$
    (d)   $Q = 0, \bar{Q} = 0$.

(5)   The outputs of a J-K bistable reverse state on each consecutive clock pulse if:
    (a)   $J = 1$ and $K = 0$
    (b)   $J = 1$ and $K = 1$
    (c)   $J = 0$ and $K = 1$
    (d)   $J = 0$ and $K = 0$.

(6)   A 4-bit binary counter is a:
    (a)   Modulo-4 counter
    (b)   Modulo-8 counter
    (c)   Modulo-16 counter
    (d)   Modulo-2 counter.

(7) The output logic states of a decade counter after the application of 10 clock pulses will be:
    (a)    0000
    (b)    1010
    (c)    1001
    (d)    1011.

(8) The propagation delay of a J-K bistable (TTL) is typically:
    (a)    20 s
    (b)    20 ms
    (c)    20 μs
    (d)    20 ns.

(9) The most likely output states of a 2-to-4 line decoder when the input lines are both at logic 0 are:
    (a)    0 0 0 0
    (b)    0 0 1 1
    (c)    1 1 1 1
    (d)    0 1 1 1.

(10) The Q outputs of an UP/DOWN counter are $Q_A = 1$, $Q_B = 1$, $Q_C = 0$ and $Q_D = 1$. The $\bar{Q}$ outputs will be:
    (a)    $\bar{Q}_A = 0$, $\bar{Q}_B = 0$, $\bar{Q}_C = 1$ and $\bar{Q}_D = 0$
    (b)    $\bar{Q}_A = 1$, $\bar{Q}_B = 1$, $\bar{Q}_C = 0$ and $\bar{Q}_D = 1$
    (c)    $\bar{Q}_A = 1$, $\bar{Q}_B = 1$, $\bar{Q}_C = 1$ and $\bar{Q}_D = 0$
    (d)    $\bar{Q}_A = 0$, $\bar{Q}_B = 0$, $\bar{Q}_C = 1$ and $\bar{Q}_D = 1$

# ANSWERS TO QUESTIONS

## ANSWERS TO QUESTIONS SET AT END OF CHAPTERS

| Chapter 1 | Chapter 2 | Chapter 3 | Chapter 4 |
|-----------|-----------|-----------|-----------|
| No.  1 (c) | No.  1 (a) | No.  1 (c) | No.  1 (c) |
| 2 (a) | 2 (d) | 2 (b) | 2 (a) |
| 3 (d) | 3 (c) | 3 (a) | 3 (a) |
| 4 (c) | 4 (a) | 4 (d) | 4 (b) |
| 5 (a) | 5 (b) | 5 (a) | 5 (c) |
| 6 (b) | | 6 (b) | 6 (a) |
| 7 (d) | | 7 (c) | 7 (c) |
| 8 (d) | | | 8 (c) |
| 9 (d) | | | 9 (a) |
| 10 (a) | | | 10 (c) |
| 11 (a) | | | 11 (c) |
| 12 (d) | | | 12 (a) |
| | | | 13 (b) |
| | | | 14 (b) |
| | | | 15 (c) |

| Chapter 5 | Chapter 6 | Chapter 7 | Chapter 8 |
|-----------|-----------|-----------|-----------|
| No.  1 (b) | No.  1 (d) | No.  1 (d) | No.  1 (d) |
| 2 (d) | 2 (b) | 2 (b) | 2 (b) |
| 3 (c) | 3 (a) | 3 (a) | 3 (c) |
| 4 (d) | 4 (c) | 4 (c) | 4 (b) |
| 5 (d) | 5 (d) | 5 (c) | 5 (b) |
| 6 (b) | 6 (a) | 6 (c) | 6 (c) |
| 7 (d) | 7 (b) | 7 (c) | 7 (a) |
| 8 (d) | 8 (b) | 8 (a) | 8 (d) |
| 9 (b) | 9 (d) | 9 (d) | 9 (d) |
| 10 (c) | 10 (a) | 10 (b) | 10 (a) |
| 11 (c) | 11 (b) | | |
| 12 (d) | 12 (c) | | |
| 13 (a) | | | |
| 14 (d) | | | |
| 15 (b) | | | |
| 16 (a) | | | |

# INDEX